THE FRONTIERS COLLECTION

THE FRONTIERS COLLECTION

Series Editors

A. C. Elitzur   Z. Merali   T. Padmanabhan   M. Schlosshauer
M. P. Silverman   J. A. Tuszynski   R. Vaas

The books in this collection are devoted to challenging and open problems at the forefront of modern science, including related philosophical debates. In contrast to typical research monographs, however, they strive to present their topics in a manner accessible also to scientifically literate non-specialists wishing to gain insight into the deeper implications and fascinating questions involved. Taken as a whole, the series reflects the need for a fundamental and interdisciplinary approach to modern science. Furthermore, it is intended to encourage active scientists in all areas to ponder over important and perhaps controversial issues beyond their own speciality. Extending from quantum physics and relativity to entropy, consciousness and complex systems—the Frontiers Collection will inspire readers to push back the frontiers of their own knowledge.

More information about this series at http://www.springer.com/series/5342

For a full list of published titles, please see back of book or springer.com/series/5342

Anthony Aguirre · Brendan Foster
Zeeya Merali
Editors

Wandering Towards a Goal

How Can Mindless Mathematical Laws Give
Rise to Aims and Intention?

 Springer

Editors
Anthony Aguirre
Physics Department
UC Santa Cruz
Santa Cruz, CA
USA

Zeeya Merali
Foundational Questions Institute
Decatur, GA
USA

Brendan Foster
Foundational Questions Institute
Decatur, GA
USA

ISSN 1612-3018   ISSN 2197-6619   (electronic)
THE FRONTIERS COLLECTION
ISBN 978-3-030-09310-5   ISBN 978-3-319-75726-1   (eBook)
https://doi.org/10.1007/978-3-319-75726-1

Printed on acid-free paper

This Springer imprint is published by the registered company Springer International Publishing AG part of Springer Nature
The registered company address is: Gewerbestrasse 11, 6330 Cham, Switzerland

Preface

This book is a collaborative project between Springer and the Foundational Questions Institute (FQXi). In keeping with both the tradition of Springer's Frontiers Collection and the mission of FQXi, it provides stimulating insights into a frontier area of science, whilst remaining accessible enough to benefit a non-specialist audience.

FQXi is an independent, non-profit organization that was founded in 2006. It aims to catalyze, support and disseminate research on questions at the foundations of physics and cosmology.

The central aim of FQXi is to fund and inspire research and innovation that is integral to a deep understanding of reality, but which may not be readily supported by conventional funding sources. Historically, physics and cosmology have offered a scientific framework for comprehending the core of reality. Many giants of modern science—such as Einstein, Bohr, Schrödinger and Heisenberg—were also passionately concerned with, and inspired by, deep philosophical nuances of the novel notions of reality they were exploring. Yet, such questions are often overlooked by traditional funding agencies.

Often, grant-making and research organizations institutionalize a pragmatic approach, primarily funding incremental investigations that use known methods and familiar conceptual frameworks, rather than the uncertain and often interdisciplinary methods required to develop and comprehend prospective revolutions in physics and cosmology. As a result, even eminent scientists can struggle to secure funding for some of the questions they find most engaging, while younger thinkers find little support, freedom or career possibilities unless they hew to such strictures.

FQXi views foundational questions not as pointless speculation or misguided effort, but as critical and essential inquiry of relevance to us all. The institute is dedicated to redressing these shortcomings by creating a vibrant, worldwide community of scientists, top thinkers and outreach specialists who tackle deep questions in physics, cosmology and related fields. FQXi is also committed to engaging with the public and communicating the implications of this foundational research for the growth of human understanding.

As part of this endeavor, FQXi organizes an annual essay contest, which is open to everyone, from professional researchers to members of the public. These contests are designed to focus minds and efforts on deep questions that could have a profound impact across multiple disciplines. The contest is judged by an expert panel, and up to twenty prizes are awarded. Each year, the contest features well over a hundred entries, stimulating ongoing online discussion for many months after the close of the contest.

We are delighted to share this collection, inspired by the 2016 contest, "Wandering Towards a Goal: How do mindless mathematical laws give rise to aims and intentions?" In line with our desire to bring foundational questions to the widest possible audience, the entries, in their original form, were written in a style that was suitable for the general public. In this book, which is aimed at an interdisciplinary scientific audience, the authors have been invited to expand upon their original essays and include technical details and discussion that may enhance their essays for a more professional readership, while remaining accessible to non-specialists in their field.

FQXi would like to thank its contest partner The Peter and Patricia Gruber Foundation. The editors are indebted to FQXi's scientific director, Max Tegmark, and managing director, Kavita Rajanna, who were instrumental in the development of the contest. We are also grateful to Angela Lahee at Springer for her guidance and support in driving this project forward.

Decatur, USA   Anthony Aguirre
2017   Brendan Foster
                                                                                     Zeeya Merali
                                                                Foundational Questions Institute
                                                                                     www.fqxi.org

Contents

Chapter 1
Introduction

Anthony Aguirre, Brendan Foster and Zeeya Merali

> If the moon, in the act of completing its eternal way around the earth, were gifted with self-consciousness, it would feel thoroughly convinced that it was traveling its way of its own accord.… So would a being, endowed with higher insight and more perfect intelligence, watching man and his doings, smile about man's illusion that he was acting according to his own free will.
>
> Albert Einstein (1931).[1]

Physicists tends to concern themselves with identifying the inanimate constituent elements of nature and using them to answer questions of what has happened, what will happen, and how things occur. At the most basic level, physics can be conceived as comprising a set of mathematical laws that enable us to make predictions about the future, or the past. They do this by specifying how a set of initial conditions drive these minute building blocks to combine, to interact, and to inexorably evolve.

However, physical reality can also be thought of in terms of the whole, rather than just the parts. We can ask *why* something happened. Is there a reason? Or is there a reason why there seems to be a reason? Many phenomena admit an alternative—and sometimes vastly superior—description in terms of goals, aims and intentions. The motion of particles through spacetime, for instance, can either

---

[1] Einstein, A. quoted in Strawson, G. "Nietzsche's Metaphysics?", in *Nietzsche on Mind and Nature,* ed: Dries, M. & Kail, P. J. E. (Oxford University Press, 2015).

A. Aguirre
UC Santa Cruz Dept of Physics, Santa Cruz, USA
e-mail: aguirre@scipp.ucsc.edu

B. Foster · Z. Merali (✉)
Foundational Questions Institute, Decatur, USA
e-mail: merali@fqxi.org

B. Foster
e-mail: foster@fqxi.org

© Springer International Publishing AG, part of Springer Nature 2018
A. Aguirre et al. (eds.), *Wandering Towards a Goal*, The Frontiers Collection,
https://doi.org/10.1007/978-3-319-75726-1_1

1

be calculated by considering the forces acting on them moment by moment, or alternatively by extremizing an action over the entire path extended through time. Many-body systems may appear hopelessly complex when thought of in terms of each of their individual constituents, yet they can be elegantly described if we minimize the energy or maximize the entropy of the system as a whole.

At even higher levels of description, living systems efficiently organize their simplest components with the intricate aims of survival, reproduction, and other biological ends; and intelligent systems can employ a panoply of physical effects to accomplish many flexibly chosen goals.

So, how can mindless mathematical laws give rise to aims and intentions? That was the question that FQXi posed in our 2016 essay contest: "Wandering Towards a Goal". We asked entrants to explain how goal-oriented systems arise and function in a world that is otherwise described in terms of goal-free mathematical evolution. Issues to consider included: how physical systems with the goal of reproduction evolved from an a-biological world; whether information processing, computation, learning, complexity thresholds, and/or departures from equilibrium might allow (or even proscribe) agency; what separates intelligent systems from those without intelligence; the relationship (if there is one) between causality and purpose; and whether goal-oriented behaviour is an accident or an imperative.

The topic proved extremely popular, drawing 219 entries—from every continent bar Antarctica. Given its tremendous scope, it also proved difficult to judge, leading to a three-way tie for first place. This volume contains all 17 of the winning essays, many of them extended and enhanced, e.g., by incorporating feedback received during the evaluation process.

In Chap. 2, joint first-place winner Larissa Albantakis argues that both biological and non-biological systems can develop goals, but they must be wired in a certain way in order to do so. She makes this case by contrasting the evolution of two types of artificial organisms, which are both controlled by small, adaptive neural networks, but which differ in connectivity. Albantakis claims that when both systems are exposed to the same environment and process information, only the integrated brain—containing elements that can causally constrain each other—will form an autonomous entity.

In his first-placed entry, Carlo Rovelli notes that naturalist accounts of human experience have been hindered by the difficulty in relating the abstract ideas of "meaning" and "intentionality" to the physical world. To remedy this, in Chap. 3, Rovelli starts from a purely physical definition of "meaningful information", combining Shannon's definition of "relative information" and Darwin's evolutionary mechanism, to explain how a general physical process can become a "signal" capable of carrying meaning.

Our remaining first-prize winner, Jochen Szangolies, identifies another problem that has hampered naturalistic descriptions of meaning and intentionality: If we postulate an internal observer appraising our mental representations, he notes, it begs the question of how to explain the workings of that internal observer, without creating an infinite regress of internal observers, like nested Russian dolls in the brain. Szangolies' solution, outlined in Chap. 4, is based on the work of von Neumann and endeavours to eliminate the distinction between the user and its representation.

Other winning entrants took a variety of approaches to identifying the key features that define agency, consciousness and the ability to pursue goals, and to explaining how they can emerge. In Chap. 5, Simon DeDeo argues that in order to be goal-oriented, a system must be able to reference itself, and proposes that such self-referencing can emerge through the process of renormalization. Using information theory, in Chap. 6, Erik P. Hoel posits that agents can causally emerge from microphysics because they have the property of being able to reproduce and maintain themselves. In Chap. 7, Sara Imari Walker also takes an information-based approach—but considers information as it is understood in the context of biological systems. This leads her to identify causal structures that give rise to the apparent goal-directedness of living systems. And, in Chap. 8, Sophia Magnusdottir sets out a way to quantify consciousness based on a system's ability to monitor and predict its environment and itself.

The role of the observer in quantum mechanics, and other scientific theories, inspired the essays in Chaps. 9–11. Dean Rickles uses this feature to scrutinise the assumption, in the essay question, that physical laws are indeed mindless. In Chap. 10, Ines Samengo describes how Darwinian evolution can lead minds to ascribe intentionality to parts of reality, in an effort to model it. And, in Chap. 11, Marc Séguin invokes QBism, a recently proposed interpretation of quantum theory that brings the observer to the fore. By combining QBism with the speculative claim that all possible realities that are allowed to exist by mathematics must be realised, Séguin defines "co-emergentism" to explain how goal-oriented agents can arise (and run essay contests).

Taking a more mathematical slant, in Chap. 12, Ian Durham focuses on a slightly different aspect of quantum theory—its apparent inherent randomness—and uses combinatorial mathematics to explain how this can generate a seemingly deterministic world. And, in Chap. 13, Noson S. Yanofsky also compares mathematics and physics. He explains how structure can be found by looking at a subset of mathematics to discover the way in which structure emerges when scientists seek symmetries in subsets of the physical world.

Some winners chose unconventional narrative forms to survey history, examine the present, and imagine the future. Rick Searle whimsically imagines a dialogue between pre-Socratic philosophers to illustrate how ancient peoples stopped attributing (potentially malevolent) goals and intentions to inanimate entities. In Chap. 14, Searle proceeds to examine whether computer programs of the present might ascribe intentionality to themselves, and investigates how future artificial intelligences may be goal-oriented. Alan M. Kadin, also identifies the focus on design or intention as pre-scientific and, in Chap. 15, he points to the work of Newton and Darwin to debunk this notion. And in Chap. 16, Tommaso Bolognesi uses a fictional conversation to illustrate how goals might arise in the context of a computer algorithm.

Our final two winners compare top-down and bottom-up descriptions of reality. In Chap. 17, Cristinel Stoica describes reality as a multiple-level pyramid, with physics at its foundation, and consciousness at its apex—and examines how lower-level rules give rise to higher-level experiences. And finally, in Chap. 18, George Ellis and

Jonathan Kopel explain how bio-molecules have evolved through natural selection to form the bridge between the levels of fundamental physics and life.

This compilation brings together the writings of physicists, philosophers, mathematicians, engineers, computer scientists, neuroscientists and more. The breadth of expertise amongst the winners was unsurprising given that the essay contest, by design, required entrants to think across traditional subject boundaries. Such an interdisciplinary approach will be essential for understanding the scientific mystery at the heart of our very being: the origin of aims, goals and intention in the physical world.

Chapter 2
A Tale of Two Animats: What Does It Take to Have Goals?

Larissa Albantakis

What does it take for a system, biological or not, to have goals? Here, this question is approached in the context of in silico artificial evolution. By examining the informational and causal properties of artificial organisms ("animats") controlled by small, adaptive neural networks (Markov Brains), this essay discusses necessary requirements for intrinsic information, autonomy, and meaning. The focus lies on comparing two types of Markov Brains that evolved in the same simple environment: one with purely feedforward connections between its elements, the other with an integrated set of elements that causally constrain each other. While both types of brains 'process' information about their environment and are equally fit, only the integrated one forms a causally autonomous entity above a background of external influences. This suggests that to assess whether goals are meaningful for a system itself, it is important to understand what the system *is*, rather than what it *does*.

2.1 Prequel

It was a dark and stormy night, when an experiment of artificial evolution was set into motion at the University of Wisconsin-Madison. Fifty independent populations of adaptive Markov Brains, each starting from a different pseudo-random seed, were released into a digital world full of dangers and rewards. Who would make it into the next generation? What would their neural networks look like after 60,000 generations of selection and mutation?

While electrical signals were flashing inside the computer, much like lightning on pre-historic earth, the scientist who, in god-like fashion, had designed the simulated

L. Albantakis (✉)
Department of Psychiatry, Wisconsin Institute for Sleep and Consciousness,
University of Wisconsin, Madison, WI, USA
e-mail: albantakis@wisc.edu

© Springer International Publishing AG, part of Springer Nature 2018
A. Aguirre et al. (eds.), *Wandering Towards a Goal*, The Frontiers Collection,
https://doi.org/10.1007/978-3-319-75726-1_2

universes and set the goals for survival, waited in suspense for the simulations to finish. What kind of creatures would emerge? ...

2.2 Introduction

Life, from a physics point of view, is often pictured as a continuous struggle of thermodynamically open systems to maintain their complexity in the face of the second law of thermodynamics—the overall increase of entropy in our universe [1–3]. The 'goal' is survival. But is our universe like a game, in which organisms, species, or life as a whole increase their score by surviving? Is there a way to win? Does life have a chance if the 'goal' of the universe is a maximum entropy state ('death')?

Maybe there is an underlying law written into the fabrics of our universe that aligns the 'goal' of life with the 'goal' of the universe. Maybe 'information' is fundamental to discover it [4] (see also Carlo Rovelli's essay contribution, Chap. 3 in this volume). Maybe all there is are various gradients, oscillations, or fluctuations. In any case, looming behind these issues, another fundamental question lingers: What does it take for a system, biological or not, to have goals?

To approach this problem with minimal confounding factors, let us construct a universe from scratch: discrete, deterministic, and designed with a simple set of predefined, built-in rules for selection. This is easily done within the realm of in silico artificial evolution. One such world is shown in Fig. 2.1a (see also [5]). In this environment, the imposed goal is to categorize blocks of different sizes into those that have to be caught ('food') and those that have to be avoided ('danger'), limiting life to the essential. Nevertheless, this task requires temporal-spatial integration of sensor inputs and internal states (memory), to produce appropriate motor responses. Fitness is measured as the number of successfully caught and avoided blocks.

Let us then populate this simulated universe with 'animats', adaptive artificial organisms, equipped with evolvable Markov Brains [5, 6]. Markov Brains are simple neural networks of generalized logic gates, whose input-output functions and connectivity are genetically encoded.

For simplicity, the Markov Brains considered here are constituted of binary, deterministic elements. Over the course of thousands of generations, the animats adapt to their task environment through cycles of fitness-based selection and (pseudo) random genetic mutation (Fig. 2.1b). One particularly simple block-categorization environment requires the animats to catch blocks of size 1 and avoid blocks of size 3 ("c1-a3") to increase their fitness.

In silico evolution experiments have the great advantage that they can easily be repeated many times, with different initial seeds. In this way, a larger portion of the 'fitness landscape', the solution space of the task environment, can be explored. In the simple c1-a3 environment, perfect solutions (100% fitness) were achieved at the

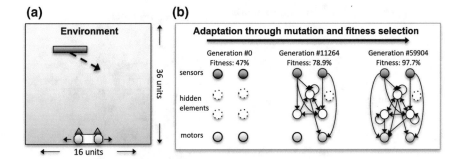

Fig. 2.1 Artificial evolution of animats controlled by Markov Brains. **a** The animat is placed in a 16 by 36 world with periodic boundaries to the left and right. An animat's sensors are activated when a block is positioned above them, regardless of distance. Blocks of different sizes fall one at a time to the right or left. Animats can move to the left or right one unit per update. **b** An animat's Markov Brain is initialized without connections between elements and adapts to the task environment through fitness-based selection and probabilistic mutation. Adapted from [7] with permission

end of 13 out of 50 evolution experiments starting from independent populations run for 60,000 generations. In the following we will take a look at the kind of creatures that evolved.

2.3 Perfect Fitness—Goal Achieved?

As in nature, various possible adaptations provide distinct solutions to the c1-a3 environment. The animats we tested in this environment [5] could develop a maximal size of 2 sensors, 2 motors, and 4 hidden elements, but were started at generation #0 without any connections between them (Fig. 2.1b). We discovered 13 out of 50 strains of animats that evolved perfect fitness, using diverse behavioral strategies, implemented by Markov Brains with different logic functions and architectures (see two examples in Fig. 2.2).

From mere observation of an animat's behavior, it is notoriously difficult to compress its behavioral strategy into a simple mechanistic description (see [8] for an example video). In some cases, an animat might first 'determine' the size and direction of the falling block and then 'follow' small blocks or 'move away' from large blocks. Such narratives, however, cannot cover all initial conditions or task solutions. How can we understand an animat and its behavior?

On the one hand, the animat's Markov Brain is deterministic, consists of at most 8 elements, and we have perfect knowledge of its logic structure. While there is no single elegant equation that captures an animat's internal dynamics, we can still describe and predict the state of its elements, how it reacts to sensor inputs, and

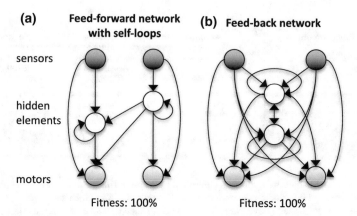

Fig. 2.2 Example network architectures of evolved Markov Brains that achieved perfect fitness in the c1-a3 block-catching task. Adapted from [5] with permission

when it activates its motors, moment by moment, for as long as we want. Think of a Markov Brain as a finite cellular automaton with inputs and outputs. No mysteries.

On the other hand, we may still aim for a comprehensive, higher-level description of the animat's behavior. One straightforward strategy is to refer to the goal of the task: "the animat tries to catch blocks of size 1 and avoid blocks of size 3". This is, after all, the rule we implemented for fitness selection. It is the animat's one and only 'purpose', programmed into its artificial universe. Note also that this description captures the animat's behavior perfectly. After all, it is—literally—determined to solve the task.

Is this top-level description in terms of goals useful and is it justified? Certainly, from an *extrinsic*, observer's perspective, it captures specific aspects of the animat's universe: the selection rule, the fact that there are blocks of size 1 and size 3, and that some of these blocks are caught by the animat and some are not, etc. But does it relate at all to *intrinsic* properties of the animat itself?

To approach this question, first, one might ask whether, where, and how much information about the environment is represented in the animat's Markov Brain. The degree to which a Markov Brain represents features of the environment might be assessed by information-theoretic means [6], for example, as the shared entropy between environment states E and internal states M, given the sensor states S:

$$R = H(E : M|S) \qquad (2.1)$$

R captures information about features of the environment encoded in the internal states of the Markov Brain beyond the information present in its sensors. Conditioning on the sensors discounts information that is directly copied from the environment at a particular time step. A simple camera would thus have zero representation, despite its capacity to make $>10^7$ bit copies of the world.

For animats adapting to the block-catching task, relevant environmental features include whether blocks are small or large, move to the left or right, etc. Indeed, representation R of these features increases, on average, over the course of evolution [6]. While this result implies that representation of environmental features, as defined above, is related to task fitness, the measure R itself does not capture whether or to what extent the identified representations actually play a *causal* role in determining an animat's behavior.[1]

Machine-learning approaches, such as decoding, provide another way to identify whether and where information about environmental variables is present in the evolved Markov Brains. Classifiers are trained to predict environmental categories from brain states—a method now frequently applied to neuro-imaging data in the neurosciences [9, 10]. Roughly, the better the prediction, the more information was available to the classifier. Just as for R, however, the fact that information about specific stimuli can be extracted from a brain's neural activity does not necessarily imply that the brain itself is 'using' this information [11].

What about our animats? As demonstrated in Fig. 2.2, the c1-a3 block-categorization task can be perfectly solved by animats with as few as 2 hidden elements. Their capacity for representation is thus bounded by 4 bits (2 hidden elements + 2 motors). Is that sufficient for a representation of the goal for survival? At least in principle, 4 binary categories could be 'encoded'. Yet, in practice, even a larger version of animats with higher capacity for representation (10 hidden elements) only achieved values on the order of $R = 0.6$ bits in a similar block-catching environment [6]. To solve this task, the animats thus do not seem to require much categorical information about the environment beyond their sensor inputs.

While this lack of representation in the animats may be due to their small size and the simplicity of the task, there is a more general problem with the type of information measures described above: the information that is quantified is, by definition, *extrinsic* information.

Any form of representation is ultimately a correlation measure between external and internal states, and requires that relevant environmental features are preselected and categorized by an independent observer (e.g. to obtain E in Eq. 2.1, or to train the decoder). As a consequence, the information about the environment *represented* in the animat's Markov Brain is meaningful for the investigator. Whether it is causally relevant, let alone meaningful, for the animat is not addressed.[2]

---

[1]Furthermore, representations of individual environmental features are typically distributed across many elements [6], and thus do no coincide with the Markov Brain's elementary (micro) logic components.

[2]Note that this holds, even if we could evaluate the correlation between internal and external variables in an observer-independent manner, except then the correlations might not even be meaningful for the investigator.

2.4 Intrinsic Information

To be causally relevant, information must be physically instantiated. For every 'bit', there must be some mechanism that is in one of two (or several) possible states, and which state it is in must matter to other mechanisms. In other words, the state must be "a difference that makes a difference" [12, 13].

More formally, a mechanism M has inputs that can influence it and outputs that are influenced by it. By being in a particular state m, M constrains the possible past states of its inputs,[3] and the possible futures states of its outputs in a specific way. How much M in state m constrains its inputs can be measured by its *cause information* (*ci*); how much it constrains its outputs is captured by its *effect information* (*ei*) [13].

An animat's Markov Brain is a set of interconnected logic elements. A mechanism M inside the Markov Brain could be one of its binary logic elements, but can in principle also be a set of several such elements.[4] In discrete dynamical systems, such as the Markov Brains, with discrete updates and states, we can quantify the cause and effect information of a mechanism M in its current state m_t within system Z as the difference D between the constrained and unconstrained probability distributions over Z's past and future states [13]:

$$ci\left(M = m_t\right) = D\left(p\left(z_{t-1}|m_t\right), p\left(z_{t-1}\right)\right) \tag{2.2}$$

$$ei\left(M = m_t\right) = D\left(p\left(z_{t+1}|m_t\right), p\left(z_{t+1}\right)\right) \tag{2.3}$$

where z_{t-1} are all possible past states of Z one update ago, and z_{t+1} all possible future states of Z at the next update. For $p\left(z_{t-1}\right)$, we assume a uniform (maximum entropy) distribution, which corresponds to perturbing Z into all possible states with equal likelihood. Using such systematic perturbations makes it possible to distinguish observed correlations from causal relations [14].[5] By evaluating a causal relationship in all possible contexts (all system states), we can obtain an objective measure of its specificity ("Does A always lead to B, or just sometimes?") [13, 15]. Likewise, we take $p\left(z_{t+1}\right)$ to be the distribution obtained by providing independent, maximum entropy inputs to each of the system's elements [13]. In this way, Eqs. 2.2 and 2.3 measure the causal specificity with which mechanism M in state m_t constrains the system's past and future states.

A system can only 'process' information to the extent that it has mechanisms to do so. All causally relevant information within a system Z is contained in the

---

[3]If M would not constrain its inputs, its state would just be a source of noise entering the system, not causal information.

[4]Sets of elements can constrain their joint inputs and outputs in a way that is irreducible to the constraints of their constituent elements taken individually [13]. The irreducible cause-effect information of a set of elements can be quantified similarly to Eqs. 2.2–2.3, by partitioning the set and measuring the distance between $p\left(z_{t\pm1}|m_t\right)$ and the distributions of the partitioned set.

[5]By contrast to the uniform, perturbed distribution, the stationary, observed distribution of system Z entails correlations due to the system's network structure which may occlude or exaggerate the causal constraints of the mechanism itself.

system's *cause-effect structure*, the set of all its mechanisms, and their cause and effect distributions $p(z_{t-1}|m_t)$ and $p(z_{t+1}|m_t)$. The cause-effect structure of a system in a state specifies the information *intrinsic* to the system, as opposed to correlations between internal and external variables. If the goals that we ascribe to a system are indeed meaningful from the intrinsic perspective of the system, they must be intrinsic information, contained in the system's cause-effect structure (if there is no mechanism for it, it does not matter to the system).

Yet, the system itself does not 'have' this intrinsic information. Just by 'processing' information, a system cannot evaluate its own constraints. This is simply because a system cannot, at the same time, have information about itself in its current state and also other possible states. Any memory the system has about its past states has to be physically instantiated in its current cause-effect structure. While a system can have mechanisms that, by being in their current state, constrain other parts of the system, these mechanisms cannot 'know' what their inputs mean.[6] In the same sense, a system of mechanisms in its current state does not 'know' about its cause-effect structure; instead, the cause-effect structure specifies what it means to *be* the system in a particular state.[7] Intrinsic meaning thus cannot arise from 'knowing', it must arise from 'being'.

What does it mean to 'be' a system, as opposed to an assembly of interacting elements, defined by an extrinsic observer? When can a system of mechanisms be considered an autonomous agent separate from its environment?

2.5 To Be or Not to Be Integrated

Living systems, or agents, more generally, are, by definition, open systems that dynamically and materially interact with their environment. For this reason, physics, as a set of mathematical laws governing dynamical evolution, does not distinguish between an agent and its environment. When a subsystem within a larger system is characterized by physical, biological, or informational means, its boundaries are typically taken for granted (see also [16]).

Let us return to the Markov Brains shown in Fig. 2.2, which evolved perfect solutions in the c1-a3 environment. Comparing the two network architectures, the Markov Brain in Fig. 2.2a has only feedforward connections between elements, while the hidden elements in Fig. 2.2b feedback to each other. Both Markov Brains 'process' information in the sense that they receive signals from the environment and react to these signals. However, the hidden elements in Fig. 2.2b constrain each

---

[6]Take a neuron that activates, for example, every time a picture of the actress Jennifer Aniston is shown [22]. All it receives as inputs is quasi-binary electrical signals from other neurons. The meaning "Jennifer Aniston" is not in the message to this neuron, or any other neuron.

[7]For example, an AND logic gate receiving 2 inputs is what it is, because it switches ON if and only if both inputs were ON. An AND gate in state ON thus constrains the past states of its input to be ON.

other, above a background of external inputs, and thus from an *integrated* system of mechanisms.

Whether and to what extent a set of elements is integrated can be determined from its cause-effect structure, using the theoretical framework of integrated information theory (IIT) [13]. A subsystem of mechanisms has integrated information $\Phi > 0$, if all of its parts constrain, and are being constrained by, other parts of the system. Every part must be a difference that makes a difference within the subsystem. Roughly, Φ quantifies the minimal intrinsic information that is lost if the subsystem is partitioned in any way. An integrated subsystem with $\Phi > 0$ has a certain amount of causal autonomy from its environment.[8] Maxima of Φ define where intrinsic causal borders emerge [17, 18]. A set of elements thus forms a causally autonomous entity if its mechanisms give rise to a cause-effect structure with maximal Φ, compared to smaller or larger overlapping sets of elements. Such a maximally integrated set of elements forms a unitary whole (it is 'one' as opposed to 'many') with intrinsic, self-defined causal borders, above a background of external interactions. By contrast, systems whose elements are connected in a purely feedforward manner have $\Phi = 0$: there is at least one part of the system that remains unconstrained by the rest. From the intrinsic perspective, then, there is no unified system, even though an external observer can treat it as one.

So far, we have considered the entire Markov Brain, including sensors, hidden elements, and motors, as the system of interest. However, the sensors only receive input from the environment, not from other elements within the system, and the motors do not output to other system elements. The whole Markov Brain is not an integrated system, and thus not an autonomous system, separate from its environment. Leaving aside the animat's 'retina' (sensors) and 'motor neurons' (motors), inside the Markov Brain in Fig. 2.2b, we find a minimal entity with $\Phi > 0$ and self-defined causal borders—a 'brain' within the Markov Brain. By contrast, all there is, in the case of Fig. 2.2a, is a cascade of switches, and any border demarcating a particular set of elements would be arbitrary.

Dynamically and functionally the two Markov Brains are very similar. However, one is an integrated, causally autonomous entity, while the other is just a set of elements performing a function. Note again that the two systems are equally 'intelligent' (if we define intelligence as task fitness). Both solve the task perfectly. Yet, from the intrinsic perspective being a causally autonomous entity makes all the difference (see here [13, 19]). But is there a practical advantage?

2.6 Advantages of Being Integrated

The cause-effect structure of a causally autonomous entity describes what it means to be that entity from its own intrinsic perspective. Each of the entity's mechanisms,

---

[8]This notion of causal autonomy applies to deterministic and probabilistic systems, to the extent that their elements constrain each other, above other background inputs, e.g. from the sensors.

in its current state, corresponds to a distinction within the entity. Being an entity for which 'light' is different from 'dark', for example, requires that the system itself, its cause-effect structure, must be different when it 'sees' light, compared to when it 'sees' dark. In this view, intrinsic meaning might be created by the specific way in which the mechanisms of an integrated entity constrain its own past and future states, and by their relations to other mechanisms within the entity.

The animat 'brain' in Fig. 2.2b, constituted of the 2 hidden elements, has at most 3 mechanisms (each element, and also both elements together, if they irreducibly constrain the system). At best, these mechanisms could specify that "something is probably this way, not that way", and "same" or "different". Will more complex environments lead to the evolution of more complex autonomous agents?

In the simple c1-a3 environment, animats with integrated brains do not seem to have an advantage over feedforward architectures. Out of the 13 strains of animats that reached perfect fitness, about half developed architectures with recurrent connections (6/13) [5]. However, in a more difficult block-catching environment, which required more internal memory ("catch size 3 and 6, avoid size 4 and 5"), the same type of animats developed more integrated architectures with higher Φ, and more mechanisms (one example architecture is shown in Fig. 2.1b). The more complex the environment, the more evolution seems to favor integrated structures.

In theory, and more so for artificial systems, being an autonomous entity is not a requirement for intelligent behavior. Any task could, in principle, be solved by a feedforward architecture given an arbitrary number of elements and updates. Nevertheless, in complex, changing environments, with a rich causal structure, where resources are limited and survival requires many mechanisms, integrated agents seem to have an evolutionary advantage [5, 20]. Under these conditions, integrated systems are more economical in terms of elements and connections, and more flexible than functionally equivalent systems with a purely feedforward architecture. Evolution should also ensure that the intrinsic cause-effect structure of an autonomous agent 'matches' the causal structure of its environment [21].

From the animats, it is still a long way towards agents with intrinsic goals and intentions. What kind of cause-effect structure is required to experience goals, and which environmental conditions could favor its evolution, remains to be determined. Integrated information theory offers a quantitative framework to address these questions.

2.7 Conclusion

Evolution did produce autonomous agents. We experience this first hand. We are also entities with the right kind of cause-effect structure to experience goals and intentions. To us, the animats appear to be agents that behave with intention. However, the reason for this lies within ourselves, not within the animats. Some of the animats even lack the conditions to be separate causal entities from their environment. Yet, observing their behavior affects *our* intrinsic mechanisms. For this reason, describing certain types of directed behaviors as goals, in the extrinsic sense, is most likely useful to us from an evolutionary perspective. While we cannot infer agency from observing apparent goal-directed behavior, by the principle of sufficient reason, something must cause this behavior (if we see an antelope running away, maybe there is a lion). On a grander scale, descriptions in terms of goals and intentions can hint at hidden gradients and selection processes in nature, and inspire new physical models.

For determining agency and intrinsic meaning in other systems, biological or not, correlations between external and internal states have proven inadequate. Being a causally autonomous entity from the intrinsic perspective requires an *integrated* cause-effect structure; merely 'processing' information does not suffice. Intrinsic goals certainly require an enormous amount of mechanisms. Finally, when physics is reduced to a description of mathematical laws that determine dynamical evolution, there seems to be no place for causality. Yet, a (counterfactual) notion of causation may be fundamental to identify agents and distinguish them from their environment.

Acknowledgements I thank Giulio Tononi for his continuing support and comments on this essay, and William Marshall, Graham Findlay, and Gabriel Heck for reading this essay and providing helpful comments. L.A. receives funding from the Templeton World Charities Foundation (Grant#TWCF0196).

References

1. Schrödinger, E.: What is Life? With Mind and Matter and Autobiographical Sketches. Cambridge University Press (1992)
2. Still, S., Sivak, D.A., Bell, A.J., Crooks, G.E.: Thermodynamics of Prediction. Phys. Rev. Lett. **109**, 120604 (2012)
3. England, J.L.: Statistical physics of self-replication. J. Chem. Phys. **139**, 121923 (2013)
4. Walker, S.I., Davies, P.C.W.: The algorithmic origins of life. J. R. Soc. Interface **10**, 20120869 (2013)
5. Albantakis, L., Hintze, A., Koch, C., Adami, C., Tononi, G.: Evolution of integrated causal structures in animats exposed to environments of increasing complexity. PLoS Comput. Biol. **10**, e1003966 (2014)
6. Marstaller, L., Hintze, A., Adami, C.: The evolution of representation in simple cognitive networks. Neural Comput. **25**, 2079–2107 (2013)
7. Albantakis, L., Tononi, G.: The intrinsic cause-effect power of discrete dynamical systems—from elementary cellular automata to adapting animats. Entropy **17**, 5472–5502 (2015)
8. Online Animat animation. http://integratedinformationtheory.org/animats.html
9. Quiroga, R.Q., Panzeri, S.: Extracting information from neuronal populations: information theory and decoding approaches. Nat. Rev. Neurosci. **10**, 173–185 (2009)
10. King, J.-R., Dehaene, S.: Characterizing the dynamics of mental representations: the temporal generalization method. Trends Cogn. Sci. **18**, 203–210 (2014)
11. Haynes, J.-D.: Decoding visual consciousness from human brain signals. Trends Cogn. Sci. **13**, 194–202 (2009)
12. Bateson, G.: Steps to an Ecology of Mind. University of Chicago Press (1972)
13. Oizumi, M., Albantakis, L., Tononi, G.: From the phenomenology to the mechanisms of consciousness: integrated information theory 3.0. PLoS Comput. Biol. **10**, e1003588 (2014)
14. Pearl, J.: Causality: models, reasoning and inference. Cambridge University Press (2000)

15. Ay, N., Polani, D.: Information Flows in Causal Networks. Adv. Complex Syst. **11**, 17–41 (2008)
16. Krakauer, D., Bertschinger, N., Olbrich, E., Ay, N., Flack, J.C.: The Information Theory of Individuality. The architecture of individuality (2014)
17. Marshall, W., Albantakis, L., Tononi, G.: Black-boxing and cause-effect power (2016). arXiv: 1608.03461
18. Marshall, W., Kim, H., Walker, S.I., Tononi, G., Albantakis, L.: How causal analysis can reveal autonomy in biological systems (2017). arXiv: 1708.07880
19. Tononi, G., Boly, M., Massimini, M., Koch, C.: Integrated information theory: from consciousness to its physical substrate. Nat. Rev. Neurosci. **17**, 450–461 (2016)
20. Albantakis, L., Tononi, G.: Fitness and neural complexity of animats exposed to environmental change. BMC Neurosci. **16**, P262 (2015)
21. Tononi, G.: Integrated information theory. Scholarpedia **10**, 4164 (2015)
22. Quiroga, R.Q., Reddy, L., Kreiman, G., Koch, C., Fried, I.: Invariant visual representation by single neurons in the human brain. Nature **435**, 1102–1107 (2005)

Chapter 3
Meaning and Intentionality = Information + Evolution

Carlo Rovelli

3.1 Introduction

There is a gap in our understanding of the world. On the one hand we have the physical universe; on the other, notions like meaning, intentionality, agency, purpose, function and similar, which we employ for things like life, humans, the economy... These notions are absent in elementary physics, and their placement into a physicalist world view is delicate [1], to the point that the existence of this gap is commonly presented as the strongest argument against naturalism.

Two historical ideas have contributed tools to bridge the gap.

The first is Darwin's theory, which offers evidence on how function and purpose can emerge from natural variability and natural selection of structures [2]. Darwin's theory provides a naturalistic account for the ubiquitous presence of function and purpose in biology. It falls sort of bridging the gap between physics and meaning, or intentionality.

The second is the notion of 'information', which is increasingly capturing the attention of scientists and philosophers. Information has been pointed out as a key element of the link between the two sides of the gap, for instance in the classic work of Fred Dretske [3].

However, the word 'information' is highly ambiguous. It is used with a variety of distinct meanings, that cover a spectrum ranging from mental and semantic ("the information stored in your USB flash drive is comprehensible") all the way down to strictly engineeristic ("the information stored in your USB flash drive is 32 Giga"). This ambiguity is a source of confusion. In Dretske's book, information is introduced on the basis of Shannon's theory [4], explicitly interpreted as a formal theory that "does not say what information is".

In this note, I make two observations. The first is that it is possible to extract from the work of Shannon a *purely physical* version of the notion of information. Shannon calls it "relative information". I keep his terminology even if the ambiguity of these

C. Rovelli (✉)
CPT, Aix-Marseille Université, Université de Toulon CNRS, 13288 Marseille, France
e-mail: rovelli.carlo@gmail.com

© Springer International Publishing AG, part of Springer Nature 2018   17
A. Aguirre et al. (eds.), *Wandering Towards a Goal*, The Frontiers Collection,
https://doi.org/10.1007/978-3-319-75726-1_3

terms risks leading to a continuation of the misunderstanding; it would probably be better to call it simply 'correlation', since this is what it ultimately is: downright crude physical correlation.

The second observation is that the combination of *this* notion with Darwin's mechanism provides the ground for a definition of meaning. More precisely, it provides the ground for the definition of a notion of "meaningful information", a notion that on the one hand is solely built on physics, on the other can underpin intentionality, meaning, purpose, and is a key ingredient for agency.

The claim here is not that the full content of what we call intentionality, meaning, purpose—say in human psychology, or linguistics—is nothing else than the *meaningful information* defined here. But it is that these notions can be built upon the notion of *meaningful information* step by step, adding the articulation proper to our neural, mental, linguistic, social, etcetera, complexity. In other words, I am not claiming to provide here the full chain from physics to mental, but rather the crucial first link of the chain.

The definition of meaningful information I give here is inspired by a simple model presented by David Wolpert and Artemy Kolchinsky [5], which I describe below. The model illustrates how two physical notions, combined, give rise to a notion we usually ascribe to the non-physical side of the gap: meaningful information.

The text is organised as follows. I start by a careful formulation of the notion of correlation (Shannon's relative information). I consider this a main motivation for this note: emphasise the commonly forgotten fact that such a purely physical definition of information exists. I then briefly recall a couple of points regarding Darwinian evolution which are relevant here, and I introduce (one of the many possible) characterisations of living beings. I then describe Wolpert's model and give explicitly the definition of meaningful information which is the main purpose of this essay. Finally, I describe how this notion might bridge the two sides of the gap. I close with a discussion of the notion of signal and with some general considerations.[1]

3.2 Relative Information

Consider physical systems $A, B, \ldots$ whose states are described by a physical variables $x, y, \ldots$, respectively. This is the standard conceptual setting of physics. For simplicity, say at first that the variables take only discrete values. Let $N_A, N_B, \ldots$ be the number of distinct values that the variables $x, y, \ldots$ can take. If there is no relation or constraint between the systems A and B, then the pair of system (A, B) can be in $N_A \times N_B$ states, one for each choice of a value for each of the two variables x and y. In

---

[1] Related ideas are developed in the teleological approach to meaning, especially work of Dretske and of Millikan. See for instance Sect. 3 in: Neander, Karen, "Teleological Theories of Mental Content",? The Stanford Encyclopedia of Philosophy? (Spring 2012 Edition), Edward N. Zalta (ed.), https:// plato.stanford.edu/archives/spr2012/entries/content-teleological/.

physics, however, there are routinely constraints between systems that make certain states impossible. Let N_{AB} be the number of allowed possibilities. Using this, we can define 'relative information' as follows.

We say that A and B 'have information about one another' if N_{AB} is strictly smaller than the product $N_A \times N_B$. We call

$$S = \log(N_A \times N_B) - \log N_{AB}, \tag{3.1}$$

where the logarithm is taken in base 2, the "relative information" that A and B have about one another. The unit of information is called 'bit'.

For instance, each end of a magnetic compass can be either a North (N) or South (S) magnetic pole, but they cannot be both N or both S. The number of possible states of each pole of the compass is 2 (either N or S), so $N_A = N_B = 2$, but the physically allowed possibilities are not $N_A \times N_B = 2 \times 2 = 4$ (NN, NS, SN, SS). Rather, they are only two (NS, SN), therefore $N_{AB} = 2$. This is dictated by the physics. Then we say that the state (N or S) of one end of the compass 'has relative information'

$$S = \log 2 + \log 2 - \log 2 = 1 \tag{3.2}$$

(that is: 1 bit) about the state of the other end. Notice that this definition captures the physical underpinning to the fact that "if we know the polarity of one pole of the compass then we also know (have information about) the polarity of the other." But the definition itself is completely physical, and makes no reference to semantics or subjectivity.

The generalisation to continuous variables is straightforward. Let P_A and P_B be the phase spaces of A and B respectively and let P_{AB} be the subspace of the Cartesian product $P_A \times P_B$ which is allowed by the constraints. Then the relative information is

$$S = \log V(P_A \times P_B) - \log V(P_{AB}) \tag{3.3}$$

whenever this is defined.[2]

Since the notion of relative information captures correlations, it extends very naturally to random variables. Two random variables x and y described by a probability distribution $p_{AB}(x, y)$ are uncorrelated if

$$p_{AB}(x, y) = \tilde{p}_{AB}(x, y) \tag{3.4}$$

---

[2]Here $V(.)$ is the Liouville volume and the difference between the two volumes can be defined as the limit of a regularisation even when the two terms individually diverge. For instance, if A and B are both free particles on a circle of of size L, constrained to be at a distance less than or equal to L/N (say by a rope tying them), then we can easily regularise the phase space volume by bounding the momenta, and we get $S = \log N$, independently from the regularisation.

where $\tilde{p}_{AB}(x, y)$ is called the marginalisation of $p_{AB}(x, y)$ and is defined as the product of the two marginal distributions

$$\tilde{p}_{AB}(x, y) = p_A(x)\, p_B(y), \tag{3.5}$$

in turn defined by

$$p_A(x) = \int p_{AB}(x, y)\, dy, \qquad p_B(y) = \int p_{AB}(x, y)\, dx. \tag{3.6}$$

Otherwise they are correlated. The amount of correlation is given by the difference between the entropies of the two distributions $p_A(x, y)$ and $\tilde{p}_A(x, y)$. The entropy of a probability distribution p being $S = \int p \log p$ on the relevant space. All integrals are taken with the Liouville measures of the corresponding phase spaces.

Correlations can exist because of physical laws or because of specific physical situations, or arrangements or mechanisms, or the past history of physical systems.

Here are few examples. The fact that the two poles of a magnet cannot have the same polarisation is excluded by one of the Maxwell equations. It is just a fact of the world. The fact that two particles tied by a rope cannot move farther apart than the length of the rope is a consequence of a direct mechanical constraint: the rope. The frequency of the light emitted by a hot piece of metal is correlated to the temperature of the metal at the moment of the emission. The direction of the photons emitted from an object is correlated to the position of the object. In this case emission is the mechanism that enforces the correlation. The world teems with correlated quantities. Relative information is, accordingly, naturally ubiquitous.

Precisely because it is purely physical and so ubiquitous, relative information is not sufficient to account for meaning. 'Meaning' must be grounded on something else, something far more specific.

3.3   Survival Advantage and Purpose

Life is a characteristic phenomenon we observe on the surface of the Earth. It is largely formed by individual organisms that interact with their environment and embody mechanisms that keep them away from thermal equilibrium using available free energy. A dead organism decays rapidly to thermal equilibrium, while an organism which is alive does not. I take this—with quite a degree of arbitrariness—as a characteristic feature of organisms that are alive.

Darwin's key discovery is that we can legitimately reverse the causal relation between the existence of the mechanism and its function. The fact that the mechanism exhibits a purpose—ultimately to maintain the organism alive and reproduce it—can be simply understood as an indirect consequence, not a cause, of its existence and its structure.

As Darwin points out in his book, the idea is ancient. It can be traced at least to Empedocles. Empedocles suggested that life on Earth may be the result of random happening of structures, all of which perish except those that happen to survive, and these are the living organisms.[3]

The idea was criticised by Aristotle, on the ground that we see organisms being born with structures already suitable for survival, and not being born at random ([6] II 8, 198b35). But shifted from the individual to the species, and coupled with the understanding of inheritance and, later, genetics, the idea has turned out to be correct. Darwin clarified the role of variability and selection in the evolution of structures and molecular biology illustrated how this may work in concrete cases. Function emerges naturally and the obvious purposes that living matter exhibits can be understood as a consequence of variability and selection. What functions is there because it functions: hence it has survived. We do not need something external to the workings of nature to account for the appearance of function and purpose.

Although variability and selection alone may account for function and purpose, they are not sufficient to account for meaning, because meaning has semantic and intentional connotations that are not a priori necessary for variability and selection. 'Meaning' must be grounded on something else.

3.4 Kolchinsky-Wolpert's Model and Meaningful Information

My aim is now to distinguish the correlations that are ubiquitous in nature from those that we count as relevant information. To this end, the key point is that surviving mechanisms survive by using correlations. This is how relevance is added to correlations.

The life of an organism progresses in a continuous exchange with the external environment. The mechanisms that lead to survival and reproduction are adapted by evolution to a certain environment. But in general environment is constantly varying, in a manner often poorly predictable. It is obviously advantageous to be appropriately correlated with the external environment, because survival probability is maximised by adopting different behaviour in different environmental conditions.

A bacterium that swims to the left when nutrients are on the left and swims to the right when nutrients are on the right prospers; a bacterium that swims at random has less chances. Therefore many bacteria we see around us are of the first kind, not of the second kind. This simple observation leads to the Kolchinsky-Wolpert model [5].

A living system A is characterised by a number of variables x_n that describe its structure. These may be numerous, but are far fewer in number than those describing the full microphysics of A (say, the exact position of each water molecule in a cell).

---

[3][There could be] "beings where it happens as if everything was organised in view of a purpose, while actually things have been structured appropriately only by chance; and the things that happen not to be organised adequately, perished, as Empedocles says" [6] II 8, 198b29).

Therefore the variables x_n are macroscopic in the sense of statistical mechanics and there is an entropy $S(x_n)$ associated to them, which counts the number of the corresponding microstates. As long as an organism is alive, $S(x_n)$ remains far lower than its thermal-equilibrium value S_{max}. This capacity of keeping itself outside of thermal equilibrium, utilising free energy, is a crucial aspect of systems that are alive. Living organisms generally have a rather sharp distinction between their state of being alive or dead, and we can represent it as a threshold S_{thr} in their entropy.

Call B the environment and let y_n denote a set of variables specifying its state. Incomplete specification of the state of the environment can be described in terms of probabilities, and therefore the evolution of the environment is itself predictable at best probabilistically (Fig. 3.1).

Consider now a specific variable x of the system A and a specific variable y of the system B in a given macroscopic state of the world. Given a value (x, y), and taking into account the probabilistic nature of evolution, at a later time t the system A will find itself in a configuration x_n with probability $p_{x,y}(x_n)$. If at time zero $p(x, y)$ is the joint probability distribution of x and y, the probability that at time t the system A will have entropy higher that the threshold is

$$P = \int dx_n dx\, dy\ p(x, y)\ p_{x,y}(x_n)\theta(S(x_n) - S_{thr}), \qquad (3.7)$$

where θ is the step function. Let us now define

$$\tilde{P} = \int dx_n dx\, dy\ \tilde{p}(x, y)\ p_{x,y}(x_n)\theta(S(x_n) - S_{thr}). \qquad (3.8)$$

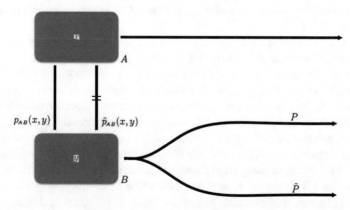

Fig. 3.1 The Kolchinsky-Wolpert model and the definition of meaningful information. If the probability of descending to thermal equilibrium $\tilde{P}$ increases when we cut the information link between A and B, then the relative information (correlation) between the variables x and y is "meaningful information"

where $\tilde{p}(x, y)$ is the marginalisation of $p(x, y)$ defined above. This is the probability of having above threshold entropy if we erase the relative information. This is Wolpert's model.

Let's define the relative information between x and y contained in $p(x, y)$ to be "directly meaningful" for B over the time span t, iff $\tilde{P}$ is different from P. And call

$$M = \tilde{P} - P \qquad (3.9)$$

the "significance" of this information. The significance of the information is its relevance for the survival, that is, its capacity of affecting the survival probability.

Furthermore, call the relative information between x and y simply "meaningful" if it is directly meaningful or if its marginalisation decreases the probability of acquiring information that can be meaningful, possibly in a different context.

Here is an example. Let B be food for a bacterium and A the bacterium, in a situation of food shortage. Let y be the location of the nutrient, for simplicity say it can be either at the left of at the right. Let x the variable that describe the internal state of the bacterium which determines the direction in which the bacterium will move. If the two variables x and y are correlated in the right manner, the bacterium reaches the food and its chances of survival are higher. Therefore the correlation between y and x is "directly meaningful" for the bacterium, according to the definition given, because marginalising $p(x, y)$, namely erasing the relative information increases the probability of starvation.

Next, consider the same case, but in a situation of food abundance. In this case the correlation between x and y has no direct effect on the survival probability, because there is no risk of starvation. Therefore the $x - y$ correlation is not directly meaningful. However, it is still (indirectly) meaningful, because it empowers the bacterium with a correlation that has a chance to affect its survival probability in another situation.

A few observations about this definition:

i. Intentionality is built into the definition. The information here is information that the system A has about the variable y of the system B. It is by definition information "about something external". It refers to a physical configuration of A (namely the value of its variable x), insofar as this variable is correlated to something external (it 'knows' something external).

ii. The definition separates correlations of two kinds: accidental correlations that are ubiquitous in nature and have no effect on living beings, no role in semantic, no use, and correlations that contribute to survival. The notion of meaningful correlation captures the fact that information can have "value" in a Darwinian sense. The value is defined here a posteriori as the increase of survival chances. It is a "value" only in the sense that it increases these chances.

iii. Obviously, not any manifestation of meaning, purpose, intentionality or value is *directly* meaningful, according to the definition above. Reading today's newspaper is not likely to directly enhance mine or my genes' survival probability. This is the sense of the distinction between 'direct' meaningful information and meaningful information. The second includes all relative information which in turn increases the probability of acquiring meaningful information. This opens the door to recursive growth of meaningful information and arbitrary increase of semantic complexity. It is this secondary recursive growth that grounds the use of meaningful information in the brain. Starting with meaningful information in the sense defined here, we get something that looks more and more like the full notions of meaning we use in various contexts, by adding articulations and moving up to contexts where there is a brain, language, society, norms...

iv. A notion of 'truth' of the information, or 'veracity' of the information, is implicitly determined by the definition given. To see this, consider the case of the bacterium and the food. The variable x of the bacterium can take two values, say L and R, where L is the variable conducting the bacterium to swim to the Left and R to the Right. Here the definition leads to the idea that R *means* "food is on the right" and L *means* "food is on the left". The variable x contains this information. If for some reason the variable x is on L but the food happens to be on the Right, then the information contained in x is "not true". This is a very indirect and in a sense deflationary notion of truth, based on the effectiveness of the consequence of holding something for true. (Approximate coarse grained knowledge is still knowledge, to the extent it is somehow effective. To fine grain it, we need additional knowledge, which is more powerful because it is more effective.) Notice that this notion of truth is very close to the one common today in the natural sciences when we say that the 'truth' of a theory is the success of its predictions. In fact, it is the same.

v. The definition of 'meaningful' considered here does not directly refer to anything mental. To have something mental you need a mind and to have a mind you need a brain, and its rich capacity of elaborating and working with information. The question addressed here is what is the physical base of the information that brains work with. The answer suggested is that it is just physical correlation between internal and external variables affecting survival either directly or, potentially, indirectly.

The idea put forward is that what grounds all this is direct meaningful information, namely *strictly physical* correlations between a living organism and the external environment that have survival and reproductive value. The semantic notions of information and meaning are ultimately tied to their Darwinian evolutionary origin. The suggestion is that the notion of meaningful information serves as a ground for the foundation of meaning. That is, it could offer the link between the purely physical world and the world of meaning, purpose, intentionality and value. It could bridge the gap.

3.5 Signals, Reduction and Modality

A signal is a physical event that conveys meaning. A ring of my phone, for instance, is a signal that *means* that somebody is calling. When I hear it, I understand its meaning and I may reach for the phone and answer.

As a purely physical event, the ring happens to physically cause a cascade of physical events, such as the vibration of air molecules, complex firing of nerves in my brain, etcetera, which can in principle be described in terms of purely physical causation. What distinguishes its being a signal, from its being a simple link in a physical causation chain?

The question becomes particularly interesting in the context of biology and especially molecular biology. Here the minute working of life is largely described in terms of signals and information carriers: DNA codes the *information* on the structure of the organism and in particular on the specific proteins that are going to be produced, RNA carries this *information* outside the nucleus, receptors on the cell surface *signal* relevant external condition by means of suitable chemical cascades. Similarly, the optical nerve exchanges *information* between the eye and the brain, the immune system receives *information* about infections, hormones *signal* to organs that it is time to do this and that, and so on, at libitum. We describe the working of life in heavily informational terms at every level. What does this mean? In which sense are these processes distinct from purely physical processes to which we do not usually employ an informational language?

I see only one possible answer. First, in all these processes the carrier of the information could be somewhat easily replaced with something else without substantially altering the overall process. The ring of my phone can be replaced by a beep, or a vibration. To decode its meaning is the process that recognises these alternatives as equivalent in some sense. We can easily imagine an alternative version of life where the meaning of two letters is swapped in the genetic code. Second, in each of these cases the information carrier is physically correlated with something else (a protein, a condition outside the cell, a visual image in the eye, an infection, a phone call...) in such a way that breaking the correlation could damage the organism to some degree. This is precisely the definition of meaningful information studied here.

I close with two general considerations.

The first is about reductionism. Reductionism is often overstated. Nature appears to be formed by a relatively simple ensemble of elementary ingredients obeying relatively elementary laws. The possible combinations of these elements, however, are stupefying in number and variety, and largely outside the possibility that *we* could compute or deduce them from nature's elementary ingredients. These combinations happen to form higher level structures that we can in part understand directly. These we call emergent. They have a level of autonomy from elementary physics in two senses: they can be studied *independently* from elementary physics, and they can be realized *in different manners* from elementary constituents, so that their elementary constituents are in a sense irrelevant to our understanding of them. Because of this, it would obviously be useless and self-defeating to try to replace all the study of nature with physics. But evidence is strong that nature is unitary and coherent, and its manifestations are—whether we understand them or not—behaviour of an underlying physical world. Thus, we study thermal phenomena in terms of entropy, chemistry in terms of chemical affinity, biology in terms functions, psychology in terms of emotions and so on. But we increase our understanding of nature when we understand how the basic concept of a science are grounded in physics, or are grounded in a science which itself is grounded in physics, as we have largely been able to do for chemical bonds or entropy. It is in this sense, and only in this sense, that I am suggesting that meaningful information could provide the link between different levels of our description of the world.

The second consideration concerns the conceptual structure on which the defini- tion of meaningful information proposed here is based. The definition has a modal core. Correlation is not defined in terms of how things *are*, but in terms of how they *could* or *could not* be. Without this, the notion of correlation cannot be constructed. The fact that something is red and something else is red, does not count as a cor- relation. What counts as a correlation is, say, if two things *can* each be of different colours, but the two *must* always be of the same colour. This requires modal language. If the world is what it is, where does modality comes from?

The question is brought forward by the fact that the definition of meaning given here is modal, but does not bear on whether this definition is genuinely physical or not. The definition is genuinely physical. It is physics itself which is heavily modal. Even without disturbing quantum theory or other aspects of modern physics, already the basic structures of classical mechanics are heavily modal. The phase space of a physical system is the list of the configurations in which the system *can* be. Physics is not a science about how the world *is*: it is a science of how the world *can* be.

There are a number of different ways of understanding what this modality means. Perhaps the simplest in physics is to rely on the empirical fact that nature realises multiple instances of the same something in time and space. All stones behave sim- ilarly when they fall and the same stone behaves similarly every time it falls. This permits us to construct a space of possibilities and then use the regularities for predic- tions. This structure can be seen as part of the elementary grammar of nature itself. And then the modality of physics and, consequently, the modality of the definition of meaning I have given are fully harmless against a serene and quiet physicalism.

But I nevertheless raise a small red flag here. Because we do not actually know the extent to which this structure is superimposed upon the elementary texture of reality by ourselves. It could well be so: the structure could be generated precisely by the structure of the very 'meaningful information' we have been concerned with here. We are undoubtably limited parts of nature, and we are so even as understanders of this same nature.

Acknowledgements I thank David Wolpert for private communications and especially Jenann Ismael for a critical reading of the article and very helpful suggestions.

References

1. Price, H.: Naturalism without Mirrors. Oxford University Press (2011)
2. Darwin, C.: On the Origin of Species. Penguin Classics (2009)
3. Dretske, F.: Knowledge and the Flow of Information. MIT Press, Cambridge, Mass (1981)
4. Shannon, C.E.: A mathematical theory of communication. Bell Syst. Tech. J. **XXVII**(3), 379 (1948)
5. Wolpert, D.H., Kolchinsky, A.: Observers as systems that acquire information to stay out of equilibrium. In: The physics of the observer Conference. Banff (2016)
6. Aristotle, Physics. In: The Works of Aristotle, vol. 1, pp. 257–355. The University of Chicago (1990)

Chapter 4
Von Neumann Minds: A Toy Model of Meaning in a Natural World

Jochen Szangolies

4.1 Introduction

Physical law and the teleonomy of living beings are difficult to reconcile: a stone rolls downhill because of the force of gravity, not because it wants to reach the bottom. Mental content exhibits the curious property of intentionality—of being directed at, concerned with, or simply about entities external to itself. Such 'aboutness' is arguably a prerequisite to goal-directed behavior; yet, following Brentano [1], it is often alleged that no physical system exhibits this kind of other-directedness.

A mark on a paper is just a mark on a paper—it is only after it is interpreted as a sign, or symbol, by an intentional mind that it comes to refer to something else. Without such interpretation, there is no reference.

But interpretation itself seems to rest on the notion of reference: when we interpret the word 'apple' as referring to an apple, a reasonable suggestion seems to be that the word causes an appropriate mental representation to be called up—that is, a certain kind of mental symbol that *refers to* said apple. We are caught in a double bind: we cannot explain reference without interpretation and we cannot explain interpretation without reference.

How, then, to reconcile the ubiquitous presence of intentionality and goal-directed behavior with blind physical law? Can we escape the double bind with a naturalist explanation, or do we have to go beyond physics, in whatever way?

In the following, I will present a 'toy model' that I argue to be capable of producing genuine meanings in a natural world. In order to introduce the model, I will, in the following section, elaborate on the nature of the double bind, introducing the *homunculus fallacy*. Following this, I will describe a model that confronts the homunculus problem head-on, eliminating the dichotomy between the representa-

J. Szangolies (✉)
Cologne, Germany
e-mail: jochen.szangolies@hhu.de; jochen.szangolies@gmx.de

© Springer International Publishing AG, part of Springer Nature 2018
A. Aguirre et al. (eds.), *Wandering Towards a Goal*, The Frontiers Collection,
https://doi.org/10.1007/978-3-319-75726-1_4

tion and its (homuncular) 'user'. To do so, I will draw an analogy to the problem of reproduction, and adapt a solution due to von Neumann [12] to the issue at hand.

4.2 The Homunculus in Locke's 'Camera Obscura'

Let us start by being exceedingly naive. We will take literally the suggestion made by Locke [7, II xi 17], that

> the understanding is not much unlike a closet wholly shut from light, with only some little openings left, to let in external visible resemblances, or ideas of things without.

The mind, then, is like a *camera obscura*: a dark room, through which light only enters via a small aperture. This produces an image of the outside world on the wall opposing the hole. In more modern terms, one could imagine an external camera, whose image is projected onto an internal screen.

It is this internal image which we will consider to form the basis of mental representation. By means of this image, actions could be planned: if the image contains an apple on a plate, one could plan to reach for and eat this apple. Thus, it seems that such an internal image suffices to implement goal-directed behavior, to plan intentional action in the outside world.

But of course, we have long since fallen prey to fallacy. When we consider how the internal image is *used* in order to formulate plans for actions in the world, we invariably postulate a *user*. This use requires recognizing the objects within the image, and representing them as apples and plates, rather than colorful blobs on a wall. But which faculty is to play the role of this user?

In attempting to give an account of the intentional machinery, we have implicitly relied on it being already present: some entity 'looks at' the internal screen, analyzes the picture, to formulate plans to enact. As the outside world is represented in this inner picture, so must the inner picture again be represented to the internal observer, the *homunculus*. This yields an infinite regression: the homunculus must contain, within itself, another homunculus accounting for its intentionality, thus yielding an infinitely nested structure of interior observers.

We thus must start out by immunizing ourselves against homunculi. To do so, we analyze the structure that underlies the notion of meaning. Meaning is a relation—but not, as one might believe at first, a two-place relation between a symbol and its referent, but a three-place relation: a symbol S means some referent R to an agent A. Thus, 'apple' (the symbol) means apple (the referent) to somebody sufficiently familiar with the English language, while it means nothing to somebody speaking only Chinese.

The source of the homunculus fallacy is glossing over whom a given symbol is supposed to have meaning to: we imagine that the internal picture is simply intrinsically meaningful, but fail to account for how this might come to be—and simply repeating this 'inner picture'-account leads to an infinite regress of internal observers.

Thus, if we intend to hold fast to a representational account of meaning, we must modify the underlying structure: replace the three-place relation by something that does not depend on external agency. Such a replacement structure was proposed by the present author in Ref. [11]; in the following, I will introduce, as well as elaborate on, the model presented therein.

4.3 Von Neumann Replicators

We have seen that the main obstacle towards finding a representational theory of meaning is the three-partite structure of reference. In order to break free of this problematic structure, we will exhibit an analogous relation, that of construction, and show how the problem arises in this case. Afterwards, we discuss the ingenious solution of von Neumann [12].

Construction is a three-partite relationship between the blueprint B, the object to be constructed O, and the constructor C. Just as a symbol S means R to the agent A, the blueprint B 'means' the object O to the constructor C. Due to this structural equivalence, we expect an analogue to the homunculus problem to surface in this context. It is not hard to see that this is, in fact, the case: consider the task of creating a self-reproducing automaton.

One possible solution is the doctrine of *preformationism*: the ancient notion that organisms contain tiny versions of the adult organisms within themselves—literal homunculi—which then grow to full size eventually. However, in order to yield a truly faithful copy of the parent organism, each of the tiny versions must, within themselves, already contain yet tinier versions, containing even smaller ones, and so on.

As with explaining meaning via reference, we seem to be stuck in an infinite regress in explaining self-reproductive capacities by means of the three-partite construction relation. Can we escape this conclusion?

Of course, self-reproduction is a fact of nature; moreover, nobody supposes that the capacity of living beings to self-reproduce depends on any mysterious, inexplicable powers. Thus, we may be confident that a solution exists.

A possibility is to just accept the homunculus regress head-on: there is, in fact, an infinite hierarchy of tiny versions stacked, in Russian doll-manner, within one another. Like the tower of homunculi gives meaning to an internal representation, this nesting enables infinite self-reproduction.

However, we do not typically believe that such infinite structures can exist in the real world. Consequently, solving both problems necessitates finding a real-world implementable replacement for the infinite structure.

A first attempt at a solution is the following. Suppose that a system is simply capable of scanning itself, producing a description that then enables it to construct an exact copy. This, at first blush, seems a very sensible solution enabling perfect self-reproduction.

However, this strategy runs into an immediate obstacle. One might anticipate this (and von Neumann did, see [12, p. 166]) due to the self-referential nature of the proposal: whenever a system scans itself, it will contain its own description as a proper part of itself; but this description then cannot be a description of the system anymore, since otherwise, it would necessarily contain a description of the system as containing a description of itself, and so on. Again, we seem to be faced with the problem of infinite regress.

Making this suggestion more concrete, Svozil [10] proves the following:

Theorem (Svozil) *In general, no complete intrinsic theory of a universal computable system can be obtained actively, i.e. by self-examination.*

The proof works by reduction to *Richard's paradox* [9]. This paradox notes that certain English expressions unambiguously name real numbers between 0 and 1— e.g., 'the ratio of a circle's circumference to its diameter minus 3'. These expressions can be brought into lexicographic order, giving expressions of the form 'the nth English expression defining a real number' a unique meaning.

Now consider the expression 'the real number whose nth decimal is given by 9 minus the nth decimal of the real number defined in the nth expression'. This defines a unique real number r. However, it is clear that this number cannot be in the original enumeration: it differs from the nth number in the nth decimal. But nevertheless, it is clearly defined by a finite English expression! Thus, we arrive at a paradox—the original list cannot be possible, after all.

Svozil then maps all possible responses of an automaton to binary strings, and shows that there cannot be a string among these capable of reproducing all others (i.e. the sought-for intrinsic theory): if that were the case, one could construct a new string via a diagonalization method that the automaton can output such that it differs from all the others in the list.

We again recognize the three-partite structure discussed previously: we have a code (the English language), a meaning (real numbers), and a decoding-mechanism (the notion of numbers defined in English)—compare this to the symbol, its meaning, and the agent to whom it has this meaning, or to the blueprint, the object to be constructed, and the constructor carrying out this construction task. Consequently, all three problems are really the same issue under different guises: positing external agency to decipher a 'code' of some sort ultimately leads to infinite regress, thanks to issues of self-reference.

Furthermore, we could again 'solve' the issue with an infinite hierarchy: call a number 'defined$_0$' if it is named by an ordinary English expression, 'defined$_1$' if it is named by an expression referring to the notion of 'definition$_0$', 'defined$_2$' if it is named by an expression referring to 'defined$_1$', and so on. Then, r would not be a 'defined' number, but rather, it would be 'defined$_1$', solving the paradox at the cost of an infinite hierarchy of definition.

However, we need not appeal to infinite constructions to circumvent this paradox. Rather, as shown by von Neumann [12], a construction is possible that splits the task into a syntactic and a semantic part—copying and interpreting the code. Since von

Neumann's original solution was geared towards the problem of self-reproduction, we will first exhibit it in this arena, before examining how to apply it to the problem of meaning.

Von Neumann's solution incorporates the following elements. First, we have a constructor C, which, equipped with a blueprint B_O of some object O, constructs that object. Schematically, we may write:

$$C + B_O \rightsquigarrow O.$$

(Note that this should not be taken to imply that C and B_O are necessarily consumed in the process—we only indicate the newly produced entity on the right side).

Furthermore, a duplicator D, which can duplicate any blueprint B_O, i.e. which performs the operation

$$D + B_O \rightsquigarrow B_O.$$

Additionally, we need a supervisor S, which activates first the duplicator, and then, the constructor, which leads to

$$S + D + C + B_O \rightsquigarrow O + B_O.$$

Reproduction then becomes possible by handing this assembly its own description, B_{SDC}:

$$S + D + C + B_{SDC} \rightsquigarrow S + D + C + B_{SDC}.$$

In self-reproduction, the blueprint is used in two *different* ways: first, it is merely regarded as a meaningless object, in order to be copied by the duplicator—only its syntactical properties are considered. Then, it is considered as a set of instructions—it is interpreted, that is, now its semantics (with respect to the constructor) are considered.

This avoids Svozil's theorem—and the homunculus problem—due to the fact that no intrinsic theory has to be obtained via self-inspection; rather, separating out syntactic and semantic aspects of the self-reproduction process enables the assembly $N = \{S, D, C, B_{SDC}\}$ to 'look at itself', rather than needing external agency. All of the elements of the three-place relation—the code, its meaning, and the agent deciphering this meaning—are now identified: the replicator N codes for itself, and reads its own code, to give rise to a copy of itself.

But von Neumann's construction enables more than mere replication. We may, for instance, change the original replicator N by adding an arbitrary pattern X to the blueprint, which yields

$$S + D + C + B_{SDCX} \rightsquigarrow S + D + C + X + B_{SDCX}.$$

That is, changing the 'genetic code' of N leads to changes of the phenotype in the next generation. Moreover, changes are hereditary—changing X to some X' will incorporate this change within the next generation. If such a replicator finds itself in competition for resources, certain changes may enable it to better acquire them, introducing a fitness differential: replicators better able to utilize an environment's resources will reproduce at a higher rate. Consequently, these replicators have the potential to adapt to an environment. As we will see, this is a key component in how genuine meanings arise.

4.4 Evolving Meaning

In the previous section, we have seen how the modification of the structure underlying replication allows us to evade the homunculus problem. A von Neumann replicator N does not have to rely on external agency in order to produce a copy of itself. Moreover, these replicators are capable of 'mutation': that is, a replicator N can code for a different one N', and thus, give rise to a modified system in the next generation. The structure here is bipartite: N interpretes itself as coding for N'. In this section, we will see that this feature may be used to produce symbols that refer to things beyond themselves—without having to introduce external agents to 'decode' them. Coming back to Locke's *camera obscura*, we will see how to produce images of the world that are capable of looking at themselves—representations that are their own users.

Von Neumann originally framed his work within the language of cellular automata (CA). A cellular automaton is, essentially, a conceptual abstraction of a grid of small machines (the cells), able to interact locally with their immediate environment—that is, changing their own state based on the states of their neighbors. These machines can be combined to form larger ones, patterns of cells in different states, much as one may combine gears, levers and pulleys into a mechanism.

In the following, we will imagine an agent whose brain is given by a cellular automaton. Since there are computationally universal CA, and neuronal networks such as ordinary brains can be simulated on a computer, such an agent is in principle capable of everything that an organism equipped with a more traditional brain can do.

Now, an analogy to the *camera obscura* would then be that the CA brain, in response to external stimuli, shows a certain pattern—an 'image' of the outside world. With such a theory, we are again chasing homunculi: in order for meaning to emerge, we need some agent external to the pattern, interpreting it as pertaining to the outside world.

But we have since seen how to exorcise the homunculus: break up the three-partite structure of reference—create mental representations (CA patterns) that are their own homunculi, using themselves as symbols. This becomes possible thanks to von Neumann's construction.

A given pattern of CA cells constitutes the 'state of mind' of our hypothetical agent. Now, consider what happens if this state of mind contains a von Neumann replicator N: the replicator 'interprets' itself, giving rise to a new copy of itself—that is, the agent's state of mind has meaning to itself. Of course, this 'meaning' is of a rather trivial, self-referential sort: all that it means is merely itself.

The key to shake the agent's mind free from empty, self-referential navel-gazing is the design's evolvability. Assume that the agent is subject to certain environmental stimuli. These will have some influence upon its CA brain: they could, for instance, set up a certain pattern of excitations—that is, cells in various states of activation. As a result, the evolution of patterns within the CA brain will be influenced by these changes, which are, in turn, due to the environmental stimuli.

Now imagine that there is a population of replicators active within the CA. Different designs will then unavoidably perform differently well in different environments—some might loose their ability to self-replicate completely; others, in contrast, might experience a boost, thus becoming more and more frequent within the population. Moreover, changes introduced within a replicator pattern may influence their replication; that is, these patterns will evolve towards a form more suited to the current conditions.

Effectively, the outside environment determines the *fitness landscape* for replicators in the CA-brain—they dictate which replicators enjoy reproductive success, and to what degree.

Consequently, confronted with a certain environmental situation, within the agent's CA brain, a replicator population suitably adapted to the CA fitness landscape will gradually become dominant.

Two key points must then be made here: first of all, through the process of mutation, the replicator comes to mean something beyond itself—it interpretes itself as a different pattern, again in a way independent of any outside agency. Second, the evolutionary process brings the replicator into ever-closer correspondence with the outside environment: the better adapted a replicator becomes to the CA conditions set up by the environment, the more it becomes a function of these conditions—and with that, of the environment.

Selection processes leave traces on those entities subject to them that are characteristic of the environment to which they are adapted. A dolphin's streamlined body attests to its living in a fluid medium; a fish's atrophied eyes bear witness to its having moved from an environment containing light to a lightless one, say due to a population becoming isolated in a cave that closed itself off. In evolutionary processes, the environment 'informs' organisms in the sense of giving shape to them.

Taken together, these two points mean that replicators in a CA-brain that is subject to environmental influences gradually come to be *about* that environment—they interpret themselves as patterns whose form comes to be ever more adapted to this environment, thus reflecting it.

Douglas Hofstadter coined the dictum that "[t]he mind is a pattern perceived by a mind" [6, p. 200]. In the above, we have presented a mechanism to realize this notion, in toy-model form: employing a replicating structure that interprets itself as something different from itself—that has meaning to itself, the way that the content of a mind has meaning to said mind.

Suppose now that the dominant replicator at a given time becomes capable of directing the agent's actions. It is immaterial here to speculate on how, exactly, this process might work—the only thing that matters is that the dominant replicator, best adapted to the CA-fitness landscape, and consequently, to the outside environment, is put into the driver's seat.

Due to the connection between a replicator and the environment, the 'active symbols' formed in this way can be employed just as homunculi might be employed to use the representation of the external world on an internal viewscreen. A symbol created in an environment containing, e.g., an apple on a plate may produce actions appropriate to the presence of that apple—i.e. cause the agent to reach for the apple and take a hearty bite.

We may imagine the symbol itself to be grabbing for the apple: just like the operator of a crane lends their intentionality to it, making it engage in purposeful behavior like lifting a load to the third floor, the symbol interpretes its own form as information about the environment, lending its agency to the agent it controls. In this sense, intentionality is contagious: intentional agents can lend their goal-directedness to larger systems they are part of.

In this way, the symbol becomes itself a kind of homunculus—albeit in a non-problematic way: while the agent's intentionality is derived from the symbol's, there is no infinite regress, due to the capacity of the symbol to interprete itself, and cause appropriate actions in the agent based on this interpretation.

Of course, the model as presented is highly speculative. There does not seem to be any experimental evidence for anything like these self-replicating configurations in real, organic brains—although theoretical proposals for self-replicating patterns of neurons do exist [5]. However, the above considerations may become more plausible upon realizing that they serve to solve further problems that otherwise seem difficult to account for.

One is the so-called *problem of error* [3]. This problem consists in the recognition that on many representational theories of mental content, it is difficult to account for erroneous judgments, e.g., for believing that something is present when it is in fact not. An example of this would be thinking that a strange person is present in a dark room, when in fact, there is only a jacket hung on the wall.

A theory on which a representation is supposed to mean whatever has caused its activation—that is, on which a mental symbol means 'apple' if it is triggered by the presence of apples—immediately falls prey to this issue.

If there is some mental symbol that is triggered whenever I see a strange person in a dark room—producing the appropriate response of being startled—, and it is triggered by a jacket, on such a theory the symbol does not mean 'strange person', but rather, 'strange person or jacket', and its being triggered is completely appropriate. But I don't find myself believing that there is a 'strange person or jacket' in the room, and having this belief confirmed by the presence of a jacket; rather, I find myself believing that there is a strange person in the room, a belief that will then, to my relief, be disconfirmed by finding that it is only a jacket.

The model as presented above accounts for this easily: while at first, a certain replicator causing a startling response was dominant within the mental population, it simply was not the most well-adapted to the actual environmental conditions, becoming eventually replaced by one fitting them better.

The other problem is connected with the question of why such a seemingly baroque scheme might develop within organisms at all. A possible reason for this is that all organisms face the challenge of coping with an open-ended environment—that is, with a possibly limitless set of environmental conditions. In artificial intelligence, this gives rise to the so-called *frame problem* [8]: a robot navigating an environment needs to possess information about this environment—roughly, about what kinds of problems to expect, and how objects behave, in order to solve them. But in the natural world, the set of object-behaviors and problems an organism might be faced with is not clearly delimited, and potentially infinite.

An evolutionary approach, in contrast, is capable of adapting to arbitrary environmental conditions. Thus, organisms possessing such a mechanism have an inherent advantage over organisms lacking it, and hence, enjoy greater reproductive fitness—leading, as a final consequence, to the existence of meaning, aboutness, and goal-directed behavior in the world.

An interesting parallel may be drawn here to the immune system: the so-called *clonal selection theory* due to Burnet et al. [2] postulates that the diversity of antibodies to combat an infection is due to a Darwinian process. In this way, the immune system does not need access to a near-limitless variety of appropriate 'responses' to all conceivable infectious agents, but rather, may evolve an appropriate response when necessary. Consequently, a strategy of using a Darwinian adaptation process in order to produce an appropriate reaction to near-limitless environmental variety already exists within nature's toolkit—and, as many examples of convergent evolution demonstrate, nature loves to recycle its solutions. (Indeed, the existence of this evolutionary immune response inspired the neuronal replicator model of Fernando et al. [5], via the *neural Darwinism* of Edelman [4]).

4.5 Conclusion

I have presented a toy model of how meaning, aboutness, and intentionality emerge in a natural world. The model's key insight is that, in order to solve the homunculus problem, the three-partite structure of reference—a symbol (or code) means its referent to an agent—must be broken up, since otherwise, we face an infinite regress of interpretational agencies.

I have argued that such a breaking up can be achieved by a mechanism analogous to von Neumann's self-replicating machines. Construction is likewise based on a three-partite structure—a blueprint becomes 'translated' into a certain object by a constructor. Von Neumann's design breaks up this structure, avoiding the infinite regress by creating objects capable of both copying and reading blueprints of themselves contained within themselves. Furthermore, the design enables evolvability: a von Neumann replicator N may construct a different object N' from the code within itself; N' then may itself be a replicator, possibly better adapted to its environment.

Translated back into the realm of symbols and its referents, N constitutes a symbol interpreting itself as N'—i.e. the symbol and the agent to whom it has meaning are identified, eliminating the homunculus, while nevertheless producing an interpretation of symbols.

The model then postulates that symbols come to 'embody' information about the outside world by an evolutionary process: environmental conditions, via the senses, set up a fitness landscape in an agent's 'brain', dictating which replicators—which symbols—enjoy the greatest reproductive success. In this way, the environment informs the replicators—differences in the environment lead to differences in the reproductive fitness of replicators.

Once a certain replicator has become dominant, it becomes capable of influencing the agent's actions, in such a way as to be consistent with the information about the environment that it embodies. Thus, the environment may include an element X; this causes a certain replicator with the trait x to become dominant; the replicator then interpretes itself, in turn causing actions in the agent appropriate to the presence of X. Just like a driver knows about the conditions on the road, and where to steer, the symbol knows about X, and causes the agent to act accordingly—it lends its intentionality to the agent, in a way that does not cause an infinite regress of intentional agents.

Furthermore, I have argued that the model makes headway in addressing the problem of misrepresentation and the frame problem. Misrepresentation is caused by a replicator becoming momentarily dominant that is not best-adapted to the environment, to then be replaced by a better-adapted one; furthermore, one can make a case that evolutionary approaches are capable of adapting to arbitrary environmental situations, thus alleviating the frame problem.

Despite these apparent successes, the model still leaves open problems. Some issues relate to a more precise formulation of the model itself—a precise formulation of how exactly the environment determines the fitness landscape is still outstanding. Additionally, it is not clear how the dominant replicator is selected in order to guide behavior.

Further questions relate to the implementation of the model in biological organisms. It is not clear whether organic brains actually support the replicating structures the model needs—that is, whether they are biologically capable of implementing the model in a sufficiently efficient way (that they are capable of implementing the model in principle is shown by the computational equivalence of cellular automata and neural networks).

Furthermore, there are issues regarding whether the model actually captures all of the phenomena associated with meaning. One question here is the generation of compound symbols: how do symbols for 'coffee' and 'mug' combine to form a symbol for 'coffee in the mug'? Or is there a separate symbol for such a compound entity? A promising direction here may be to again look to well-established biology as an inspiration: after all, we readily know examples of 'compound replicators'—namely, multicellular beings. Might such a strategy also work for symbol composition?

References

1. Brentano, F.: Psychologie vom Empirischen Standpunkte. Duncker & Humbblot, Berlin (1874)
2. Burnet, S.F.M., et al.: The Clonal Selection Theory of Acquired Immunity. Vanderbilt University Press, Nashville (1959)
3. Dretske, F.: Knowledge and the Flow of Information. MIT Press, Cambridge (1981)
4. Edelman, G.M.: Neural Darwinism: The Theory of Neuronal Group Selection. Basic Books, New York (1987)
5. Fernando, C., Karishma, K.K., Szathmary, E.: Copying and evolution of neuronal topology. PloS One **3**(11), e3775 (2008)
6. Hofstadter, D.R., Dennett, D.C.: The Mind's I: Fantasies and Reections on Self and Soul. Bantam Books, New York (1981)
7. Locke, J.: An Essaý Concerning Humane Understanding. London (1690) (T. Basset, E. Mory)
8. McCarthy, J., Hayes, P.J.: Some philosophical problems from the standpoint of artificial intelligence. Read. Artif. Intell. 431–450 (1969)
9. Richard, J.: Les principes des mathématiques et le probleme des ensembles. Revue generale des sciences pures et appliquées **16**(541), 295–6 (1905)
10. Svozil, K.: Randomness and Undecidability in Physics. World Scientific, Singapore (1993)
11. Szangolies, J.: Von neumann minds: intentional automata. Mind Matter **13**(2), 169–191 (2015)
12. von Neumann, J.: Theory of self-reproducing automata. In: Arthur, W. (ed.) Burke. University of Illinois Press, Champaign (1966)

Chapter 5
Origin Gaps and the Eternal Sunshine of the Second-Order Pendulum

Simon DeDeo

> How happy is the blameless vestal's lot! The world forgetting, by the world forgot. Eternal sunshine of the spotless mind! Each pray'r accepted, and each wish resign'd
> — Alexander Pope, "Eloisa to Abelard"[1]

The world we see, and the worlds we infer from the laws of physics, seem completely distinct. At the blackboard, I infer that a thin skein of gas will coalesce into objects such as stars and galaxies. With a few more assumptions I predict the range of masses that those stars should have, beginning from an account of initial quantum fluctuations. Today, it's considered a reasonable research goal to reduce even that story, of the wrinkles in spacetime that seeded Andromeda, to the first principles of basic physics: Hawking radiation at a horizon, the quantum statistics of a multiverse.

If, however, I try to infer the existence of the blackboard itself, and the existence of people who write on it and themselves infer, I am stuck. I find myself unable to predict the spectrum of desires and goals that evolution can produce, let alone the ones that arise, apparently spontaneously, from the depths of my own mind. The utter failure of otherwise reliable tools to generalize to this new domain is one that many scientists experience when they cross between fields. Not just scientists: as Sherry Turkle pointed out, even young children experience it, when confronted by electronic toys. There is something about the experience of life (or life's substrate,

---

[1] http://bit.ly/2BQganC.

S. DeDeo (✉)
Department of Social and Decision Sciences,
Carnegie Mellon University, Pittsburgh, PA, USA
e-mail: sdedeo@andrew.cmu.edu; simon.dedeo@gmail.com
URL: http://santafe.edu/~simon

S. DeDeo
The Santa Fe Institute, Santa Fe, NM, USA

© Springer International Publishing AG, part of Springer Nature 2018
A. Aguirre et al. (eds.), *Wandering Towards a Goal*, The Frontiers Collection,
https://doi.org/10.1007/978-3-319-75726-1_5

computation) that goes beyond purely physical mechanisms they're used to seeing in other toys. A child faced with an apparently living machine looks in the battery compartment to see what powers it [1]. Whether it is felt by an adult scientist at the blackboard, or a child with a toy robot, it is at heart an experience of the gap between the purposeful world of human life and the aimless one of stars. Our tools can not make the leap.

Our tools do, of course, work if we are allowed to assume the existence of meaning-making beings to begin with. Fluid dynamics can describe the flow of traffic through my city, while variants on the Ising model allows me to predict the racial segregation I see as I pass through it, and further generalizations get us off on the right foot for thinking about how my messily-wired brain might learn and remember and experience at all.

Yet no matter how well we do once meaning-making beings are taken as a given, we stumble when we are asked to predict their very being at all. It is *this* gap, the inability to leap from one side to the other, that begs explanation, and I refer to it as the Origin Gap because it is familiar to those working in the "origin" fields: the origin of society, the origin of consciousness and meaning, the origin of life. It is the gap that gives those fields a very different flavor from the sciences of their mature subjects. Origin of society looks very different from social science and anthropology; origin of consciousness looks very different from psychology; origin of life looks very different from biology.

The gap, I claim, is understandable, even (one might say) predictable. In this essay, I'll first show that the existence of the gap is the consequence of a basic pair of facts in the theory of computation. Second, that particular aspects of the laws of physics make it very likely that in the evolution of the universe, such gap will naturally appear. Taken together, these facts explain how "mindless" laws lead to the emergence of new realms of intentional behavior. At the heart of this essay's explanation of the gap will be that the kind of intention, aim, and meaning we really care about also has the capacity to refer to itself.

5.1 The Mathematics of the Gap

From the mathematical point of view, the origin gap begins with the fact that

1. it is easy to describe everything.
2. it is much harder to describe one thing.

This has a counterintuitive feel to it. We began, both as individuals and as a species, by describing particular things (that big mountain, this frozen river, that tall woman, this cold morning). It therefore feels as if this task must be easier than the more elaborate habits of generalization, abstraction, the tools of set theory, category theory …

Yet when we make this leap, we forget how many millions of years of evolution went into teaching us how to produce these descriptions. What it means to be one thing rather than two, the identification of useful boundaries or persistent patterns, what it means for an argument to be valid: each is a question subject to endless debate. We see this in the history of philosophy, but a more contemporary example comes from the history of Artificial Intelligence (AI). AI gave a name to the feeling that rules of description could never be exhaustively specified. They called it the "frame problem", and advanced societies across the globe dumped literally billions of dollars into solving it. Until, that is, they discovered that the quickest way to solve the problem of describing something was to avoid specifying the rules at all.

Rather than define in computer code a beautiful sunset, or a valid argument, researchers now build learning machines that watch and copy human response. Don't describe a cat to a computer; have it learn what a cat is from the pictures we take to celebrate them on the internet. In this way, the code can rely on the accumulated wisdom of evolution. Which is only natural, since (of course) a computer is build by evolved creatures to serve their needs.

In as much as our lives are dominated by artificial intelligence we have, for now, given up on describing things. But it turns out that to describe *everything*, by contrast, is simple. It only needs to occur to you to do something so trivial as to try. Borges did so in his short story *The Library of Babel*, where he imagined a series of interconnected hexagonal rooms, walled by shelves and stacked with books, and each book containing the letters of the alphabet, spaces, commas, and periods in different orders.

How much is contained in everything! Of course, in Borges' library there are an overwhelming number of nonsensical books, cats typing on keyboards, but also (again, of course) the complete works of Shakespeare, as well as every variation on those works, and every possible edition with typographical errors, and (as Borges might have gone on) the plays that Shakespeare might have written were he really Francis Bacon, or Elizabeth the First, or an alien from Mars, as well as all the incorrect extrapolations of those conjectures, and so on to the limit of one's imagination, and (then) beyond.

Imagine that we have instant access to the text of any book. It's simple to find all the books that include the word "Shakespeare": just send your robot out to search book by book and return the ones that contain that string of letters. Of course, it will also recover nonsense books, books full of jumbled letters that happen, once in awhile, to spell the name: "...casa,cWas,,, qwh g Shakespeare acqq CO..."

Here's a harder problem: how to locate the books on Shakespeare that make sense? Give instructions to the robot to gather them together. Or, imagine the layout of the Borges library as a wireframe image on your computer screen, and the rules of shelving to hand. Outline, or click with a mouse, the shelves to pull.

Under some very basic assumptions (which we'll address below), the strange thing about this more complicated query that it can imagined, but not actually made. Like the idea of squaring the circle, of producing using straight-edge and compass a square whose area is equal to a given circle, it seems that it should be possible. And yet it is not: the shape your mouse carves out, although imaginable in each fragment—"this book, not that"—is an impossible shape, a shape impossible to define and therefore to draw. In its infinitely detailed structure, it is at each scale completely unrelated to the scale above.

Computer scientists usually introduce these shapes in a very different fashion: by describing things that are capable of self-reference, and then by showing how questions about these self-referring things, though well-phrased, have no answers. Consider, for example, a game two mathematicians might play: name the number. "The smallest prime greater than twelve", for example, names the number thirteen. "Two to the power of fifty" names a much larger number, something just a bit bigger than 100 trillion. There are better and worse ways to name something ("one plus one plus one plus..." is a poor way to start naming a number larger than ten thousand, say), and you might imagine mathematicians competing to name something in the most efficient, the shortest, way.

A classic example of how the game goes wrong was provided, appropriately enough given our introduction of the Borgesian Babel Library, by an Oxford librarian, G.G. Berry, who asks us to consider the following sentence:

The smallest number that can not be named in less than a thousand words.

Such a sentence has a twisty logic to it: whatever it names, it certainly names in less than a thousand words. And yet whatever it purports to name must be something that actually requires the far larger sentence. The resolution of a paradox like this is not to reject the sentence, but to rule out the possibility of the efficient mathematician, a person (or machine) that finds the shortest description of any number. The problem that Berry's paradox reveals is that problem with certain kinds of systems that can refer to themselves: Berry's paradoxical sentence refers (implicitly) to the very practice it enacts, that of finding short descriptions. It's an easy matter for the impossibilities implied by self-reference in the Berry case to lead to the problems of locating books on Shakespeare in the Borgesian library.

The existence of such shapes (or the non-existence of the rules of their construction) seems counter-intuitive at first, because it is the nature of human beings to ask for things that are possible. "Bring me all the sugar in the kitchen"; "Find me all the students in the engineering department". We are not used to asking questions that have no answer.

Yet for it to happen all we need is that any description of what it means to be a sensical book on Shakespeare requires more than just pattern-matching (e.g., the presence or absence of the word "Shakespeare"). Impossible questions emerge when they become about pattern-processing, pattern manipulation, pattern computation. Something sophisticated enough, in particular to allow us to have something operate on a description of itself.

This is, of course, exactly what takes place. If we read a book on Shakespeare we do more than count words and match them to lists. We think about those words, the combinations they fall in, and what one combination means for another. We reason about a passage, follow its arguments and conjecture counterarguments. And when we give ourselves, or a machine, that power, we become fundamentally limited in the questions we can ask and answer about what we, or it, is going to do. It becomes impossible, even, to draw outlines around the behaviors we do, or do not, expect. In contrast to the condensation of gas into stars, we can not derive, ahead of time, the space of books that scholars will write about Shakespeare.

This is why the origin problems are hard. The things whose origins most intrigue us are also the points at which systems gain new powers of self-reference. And these moments lead to new categories, new phenomena, that we can literally not predict ahead of time. Once we have an example, we can ask questions about it, do science on it, just as we can take any particular volume from the Borgesian library and read it. But to begin with the space of all possible things that can happen, and then to draw the outlines of what we expect to see on the basis of a self-referential process, is something else altogether.

I'm hardly the first to draw attention to the importance of self-reference for the problems of life. Sara Walker and Paul Davies have pointed to the self-referential features at the heart of the origin of life problem [2]. Stuart Kauffman puts self-reference at the heart of both biological and social evolution, and in places conjectures explicitly Gödelian arguments [3]. My own work, and that of my collaborators, on social behavior suggests that social feedback, the most primitive form of self-reference and something we see in the birds just as much as the primates [4], is at the origin of major transitions in political order [5].

The gap between physics and the meaningful experiences we associate with life thus turns out to have an unexpectedly mathematical feel. The emergence of meaningful experiences is associated with the emerge of new forms of self-reference, but questions about the basic properties of self-referential systems are (on pain of logical inconsistency) impossible to answer in the complete and general fashion we expect from derivations in the physical sciences.

Asked to sketch out the consequences of a new self-referential phenomenon—say, an organic polymer than can refer to, modify, and reproduce itself—we stumble, because the very question is unanswerable. Given a particular example (the replication machinery of the bacteria *E. coli*) we can do a great deal of science. But to delineate all the life this makes possible is equivalent to picking books of Borges' library.

Before moving to the next part of this essay's argument, the physics of self-reference, it's worth pointing to the leap that's implicitly being made here. The Earth, and everything on it, is finite in nature: only so many things will ever happen. If the holographic principle is true, we may even be able to compute the total entropy contained within the boundary of our planet's world line. This enables purists to object to the arguments I've made here, because Gödelian impossibility theorems usually require an infinity somewhere. Explicitly, the things that self-reference makes impossible are those that are required to apply to everything in the domain in question: every number in the set of integers, every program that could be written and how it behaves on every set of inputs. All the numbers involved (the size of the Borgesian library, whose books are of limited size; indeed, the number of behaviorally-distinguishable possible configurations of the human mind) are not infinite, but rather simply very, very, very large. This means it is possible to tell your assistant what to do: you could, for example, go out yourself, read all the books shorter than a certain length, and give him a list. Irritating as these list-based solutions are, it's rather hard to rule them out. They're clearly unsatisfactory, because they somehow presume the answers are already to hand, a little like giving someone a grammar of the English language that simply lists all sentences shorter than ten thousand words. We want something that summarizes, compresses, or otherwise gives us rule-based insight.

When infinities go away, however, it can become possible to approximate the things we want without falling victim to paradox. We often want to talk about "shortest descriptions", for example, even when their kind of twisty self-reference puts us in the cross-hairs of Berry's paradox. In a 2014 paper Scott Aaronson, Sean Carroll, and Lauren Ouellette squeezed around it by using the file compression program `gzip` [6]! It's a clever idea, and (in my opinion) an excellent way to probe the problem they have at hand, but it's not going to work in all situations: I shouldn't try to judge the complexity of a student's reasoning by looking at the filesize of a `gzipped` version of her text. While we have heuristics and good ideas in some situations, we don't yet have a good handle on how an impossible problem "degrades" into a solvable one more generally. It's likely that the theory of computational complexity will play a role (see Ref. [7] for a philosophical overview).

Noam Chomsky confronted this problem head on in *Aspects of the Theory of Syntax* [8], where he distinguished performance (what we say) and competence (grasping the rules of what we say), and introduced to linguistics the idea that good rules, the kind of rules we want, are "generative". Something like Chomsky's competence-performance distinction, and insistence on the creation of generalizable rules rather than the creation of endless descriptive categories, is part of the story.

5.2   The Physics of the Gap

It is one thing to ascribe the gap to the emergence of new systems of self-reference. But why should self-reference come into being at all? At the heart of self-reference is the existence of a memory device, and something that can navigate it in a "sufficiently sophisticated" fashion. Smith and Száthmary's famous 1997 piece [9], on the major transitions in the biological record, recognizes this implicitly, placing the discovery of new information processing and recording mechanisms at the center of each transition. Social scientists [10], scholars of "deep history" [11], and cognitive scientists [12] each draw attention to new institutions, like cities, or new cultural practices, such as writing or social hierarchy, or even new abilities from physiology itself, such as genes for speech and syntactic processing, that enhance the ways in which we can remember and transform what we remember. Major leaps occur when something previously forgettable, lost to noise, finds a means to be recorded, translated into a referential form, processed and combined with others. When social debts become stories told around a campfire—or transform into money and markets [13]—we see not just an augmentation of life as it was known, but the unpredictable creation of entirely new forms of being.

Each of these moments is a shift in the nature of the world, and a clear topic of scientific study. Whether we study the details of its emergence, or the patterns it displays that generalize beyond its historical context, any one of them is the task of a lifetime. But what makes memory, and self-reference, possible at all?

Strangely enough, it's not baked in to the fundamental laws of physics, a fact that was driven home to me early in my career, at the University of Chicago, when I worked with Dimitrios Psaltis, and Alan Cooney, physicists at the University of Arizona. We were puzzling over a strange class of models in fundamental physics. Despite their mathematical coherence, they were, at heart, unstable: any universe that obeyed their laws would sooner or later explode, everywhere and instantaneously, into fountains of energy with no apparent end, as if slipping off the top of a hill that had no bottom.

What made them unstable was how they handled time. In the physics you encounter in high-school it's crucial that Newton's laws of motion talk about the relationship between force and acceleration: $F = ma$, force is mass times acceleration, or perhaps more easily, $a = F/m$, the acceleration you experience is the force applied to you, divided by your mass. Acceleration is connected to the passage of time; it's how fast your velocity is changing, or, more formally, the "second derivative of position with time". Newton's laws then connect forces you might experience to a phenomenon we call gravity: objects create a gravitational field, and at each point that field subjects objects to a certain amount of force. Other laws talk about other sources of force: electrical, or magnetic, for example. All connect back to acceleration, the change of velocity with time.

This is all awesome and highly addictive to talk about if you have a certain bent of mind, but one of the basic facts about these laws is that you never see anything with more than two derivatives in the fundamental equations. When you write them down, you only ever talk about (1) a basic set of quantities, say, position, gravitational field, etc.; (2) how these quantities change with time; and (3) sometimes, how these changes in time change with time. If you have a theory where higher derivatives enter in, where you talk about changes in changes in changes, then the theory becomes unstable in some really uncomfortable ways, leading to things like spontaneous infinite accelerations which you never observe (or really could imagine observing) and that would really ruin your day if you did. This has been known since the 1850s, when the Russian physicist Mikhail Vasilevich Ostrogradsky published what is now called "Ostrogradsky's Theorem" in the journal of the Academy of St. Petersburg. At the end of this essay I provide an afterword that gives the underlying physical intuition for why this is true, through an economic analogy to a shift in marginal costs. A technical introduction can be found in the account by physicist Woodard [14], while Dimitrios, Alan, and I were working on how to "cure" these instabilities in certain limited regimes; you can find our answers (and further references) in a series of papers we wrote together [15, 16].

Ostrogradsky's theorem sounded just fine to me, until I remembered something from my high school physics teacher. The change in acceleration, the *third* time derivative of position, has its own name—"jerk". Jerk is what you experience when an elevator starts up. When it's moving at a constant velocity, you feel nothing. When it's accelerating, you feel heavier (if you're going up), or lighter (if you're going down). But when it switches from not accelerating to accelerating, or vice versa, you experience a sudden change in your weight. You're experiencing the elevator jerking you up, or the pit of your stomach dropping out when it descends.

The fact that I experience jerk is very strange. Am I not a creature that lives in the physical world? Am I not forced to obey the laws of physics? And don't I know, from a bit of mathematics, that the laws of physics only deal with quantities with two time derivatives or fewer, or risk being violently unstable if they don't? But if all that's true, how can jerk, a third-order quantity, play any causal role in my life, such as causing me to say "oof", or making me feel queasy, when the elevator moves? How can my psychological laws obey equations that are ruled out as physical laws?

I remember a spooky feeling when I put this argument together, and for a brief moment wondering if this proved the existence of a separate set of psychological laws beyond or parallel to physics. The answer turns out to be a bit simpler, if no less intriguing. The instabilities that emerge for theories with higher-order derivatives are real, and barriers to them being basic laws of the universe are real as well. But there's nothing that prevents them holding for a while, in limited ways, so that the instabilities don't have time to emerge.

And that's the reason I can feel the jerk. I have a brain that senses acceleration. It's possible for that sense to rely directly on fundamental laws (it doesn't, actually, but it could). But in order to report the sensation of jerk to my higher-order reason, my brain has to go beyond fundamental physics. It has to use memory to store one sensation at one time and compare it, through some wetware neural comparison device, to a sensation at a later time. Similarly, I can measure the acceleration that my car undergoes by hanging a pendulum from the ceiling and seeing where it points, leveraging a little bit of fundamental physics. But to measure jerk, I have to videotape the pendulum, and compare its location at two different times. There's no "jerk pendulum" I can build that relies directly on the basic laws of physics that apply everywhere and for all time. The fundamental laws are forgetful, the "blameless vestals" of the Alexander Pope quotation that begins this essay.

It's strange to think that a visceral and immediate feeling, like the drop you feel in the pit of your stomach when the elevator descends, is an experience filtered through a skein of memories. These memories present what is actually a processed and interpreted feature of the world as if it were a brute physical fact. Yet it so turns out that some things, like "force", are truly fundamental constituents of our universe, while others, like "jerk", are derived and emergent.

Jerk gets into the physical world through memory, but it's hardly the most impressive feat of memory we do. A man descending a New York City skyscraper is in the presence of far greater feats of memory and processing than just what travels down his vagus nerve. Yet jerk also gives us a clue to how those far more sophisticated memories might have gotten going. The experience of jerk is an atavism of a far more primal event, one that began well before there were brains to feel it.

This is because, while (to the best of our knowledge) higher order "memory terms" like jerk are forbidden from playing a role in fundamental laws, they do emerge in an unexpected fashion. We rarely perceive the world at its finest grain, in all of its fluctuating detail, at the assembly code level, you might say as a computer programmer. We see, instead, averages: not everything that happens in a single patch, but a coarse-grained summary of it, a blurring of details as if the lens was smeared with vaseline. To give a full account of the role of that averaging, or coarse-graining, in the physical sciences would take us very far afield, but also (many now believe) into some of the best mysteries we have to hand, including an explanation of the second law of thermodynamics and the decoherence, or collapse, of the quantum-mechanical wavefunction.

Here we care about coarse-graining because, by averaging together nearby points, it introduces the possibility of inducing physical laws that (in contrast to their forgetful fundamental cousins) do have memory. When we smooth out the world, when we average out some of the small-scale bumps and fluctuations, we produce a new description of it. The laws that govern those coarse-grained descriptions, in contrast to the ones that applied at the shorter distances and for the finer details, can have memories, can include higher derivatives. They may, in certain cases, be unstable, but this is no longer an existential threat: it just means that, occasionally, the coarse-grained description will fail. The fine-grained details will emerge with a vengeance, ruining the predictive power of the theory. You'll be reminded of the limits of your knowledge, but the universe will not catch fire.

The technical term that physicists use for this is renormalization. Physicists use it for all sorts of problems, and call the theories that emerge for coarse-grained systems "effective theories". My colleagues and I have thought about them for a long time, as both a fact of life for deriving one scale from another (social behavior, say, from individual cognition), and a metaphor to help explain why biology differs so much from biochemistry, and why averaging-out might not just be a good idea, but might make new forms of society possible [4, 5, 17–19].

If you're a computer scientist, you might say that while the universe runs in assembly code, the coarse-grained version runs in LISP. Here, coarse graining gives the possibility of memory and—with some interesting dynamics for how those memories inter-relate—the self-reference that makes certain features of the future logically unpredictable based on what came before. The memories we have now, biochemical, electronic, on pen and paper and in the cloud, are far more complex than the ones than appear in a physicist's coarse-graining prescription.

You get a great deal from the averaging-out a cell wall allows you to do, another boost from the ways in which neurons average out the data from your eye, and another from how a story you tell summarizes the history of your tribe. No essay can derive the biochemical story, or the cognitive one, or the social one. Here we point to a crucial moment where they all begin: not in the perception of detail, but in its selective destruction and lossy compression. The arguments of this essay suggest that averaging-out may have been the first source of memory, and thus self-reference, in the history of the universe. Perhaps that happened first in biochemistry; perhaps it had an even earlier start.

5.3   Learning from the Gaps

The leaps the universe has made, from non-life to life, chemical reaction to mind, individualism to society, aimless to aim-ful depend in a basic way on how new features of the world—physical features, biological features, social features—become available for feedback and self-reference. If this essay is correct, then it is those self-referential features that, in creating predictive gaps, attract our curiosity. And it is those features that, at the same time, make the problems so hard. You might say we're constantly nerd-swiped by the origin gaps [20].

Though I've focused on the primordial scene, the origin of memory, and located it in the coarse-graining of fundamental theories, I've also suggested that this coarse-graining process might be something worth attending to at later stages as well. This suggests an intriguing possibility: that there are more stages yet to come, new accelerations and ways for us to reflect upon ourselves, and (in doing so) to create new forms of life. It's natural, at this cultural moment, to look to the world of artificial intelligence and to ask what our machines will do for—or to—us. As we create machines with inconceivably greater powers to reflect, we may be setting in motion a process that will leave behind, for future millennia, a new origin problem to solve.

5.4   Afterword: The Economics of Physical Law

In the classical world, i.e., the world before we consider quantum effects, we describe the behavior of everything from planets to beachballs by talking about how they respond to the forces placed on them, and how they might create forces that others respond to in their turn. A basic feature of these laws, as far as we understand them, is that the only things we need to know are the positions of the particles, their velocities, and how their velocities change in time (their accelerations). A planet's gravitational field tugs on a beachball, causing it to descend; the effect of the planet on the ball can be summarized, without loss, by talking about how the ball accelerates in time. The position of the beachball (how far away it is from the center of the each) dictates its change in velocity with time (acceleration). In more complicated situations, such as magnetism, the acceleration of a particle might depend on its velocity as well as its position. But none of the laws we know of tell us, for example, give an independent role to (say) jerk, the change in acceleration with time.

A higher-derivative theory, by contrast, is one where facts about these more derived changes do have an independent causal power over the system's evolution. We can write them down, if we like, but when we examine the predictions they make, we find that they not only do not describe the world we know, but in important senses they can not describe anything that remotely looks like the world we know. We've known this since Mikhail Vasilevich Ostrogradsky published what is now called "Ostrogradsky's Theorem" in the journal of the Academy of St. Petersburg, in 1850.

Richard Woodard has an excellent, technical introduction to Ostrogradsky's theorem on Scholarpedia [21].

Here, I'll try to give an intuitive introduction to the physics behind it, by drawing a parallel to economics. Physics and economics have long travelled in parallel. Students of the great physicist, and modern interpreter of thermodynamics, Josiah Willard Gibbs, went on to define 20th Century economics, at least for the Americans: Paul Samuelson was a direct descendent, who not only won the Nobel Prize in economics, but trained generations of policy-makers to come through his 1948 textbook Economics: An Introductory Analysis. But the parallels go back further in time, and in previous centuries yet it was not uncommon for someone today known for contributions to physics to have also been intrigued by, and often an originator of, basic concepts in economics. Let's go back to that tradition to see what happens.

Since the late 18th Century, physicists have defined theories by describing how particles move and arrange themselves in space so as to minimize a particular quantity. The quantity, called the *Lagrangian*, attaches a value to every possible arrangement of particles in space and (crucially) to their velocities as well. Think of the Lagrangian as defining a cost that the particles pay for being in a certain place with a certain velocity. As these particles move through space, passing from one arrangement to another (or staying in the same configuration for a while) they run up a bill. Some configurations of course, are more costly than others; some are costly, but enable the particle to get to less costly configurations later.

To figure out how particles set in motion at one time (say, 9 am) will move and interact with each other, ending up in a new configuration at (say) 10 am, we consider all possible paths the particles might have taken to get from the arrangement at 9 am to the arrangement in question at 10 am. Each path runs up a bill, and the actual path that particle takes is that which runs up the smallest Lagrangian bill. Some paths are absurd, with particles stopping and starting at random, accelerating to vast speeds and just as quickly coming to a halt; these end up running up very large bills. The ones that rack up smaller bills are close to the paths the particles actually take, and the very smallest bill is the path the particles do take. (Hidden in here is the secret for generalizing to quantum mechanics: we now allow particles to stray from the optimal paths, penalizing them the further they stray, and indeed, all the arguments we're about to make remain valid for the quantum world).

One of the puzzles of the Lagrangian formulation is that it's hard to think of this path selection as a causal story, in the usual sense: as a particle moves in space, it may be able to achieve a minimal bill by temporarily paying large costs "in anticipation" of reduced costs later in time—somewhat like a young financier living in New York City in his early twenties, only to move to Connecticut in mid-life. But in another sense, causality is preserved: only facts about the Lagrangian bill matter, and the only way to manipulate the particle is to alter, or add, terms in the Lagrangian. The particle doesn't really anticipate and think through the consequences of its behavior, although in the quantum mechanical formulation, you can think about it as trying a bunch of different paths in parallel universes.

In any case, ordinary Lagrangians only bill particles in terms of their positions and velocities: informally, it costs something to climb a hill, and it costs something to be fast. All paths can be defined in terms of their position and velocity alone; you can measure the implied acceleration, of course, at any point, but accelerations only matter to the extent to which they end up affecting the particle's velocity or position. Nothing else is billed: it comes for free, like the bread before the appetizer.

We can ask how much *more* it costs to be fast, given a little boost in speed (i.e., when we take the derivative of the Lagrangian with respect to the velocity). If I'm already going 50 miles/h, how much more does it cost to go to 51? If you're an economist, you can think of this as the *marginal cost*: if I'm already making fifty thousand widgets, how much more does it cost to make the next?

Let's persist with the economic analogy. In general, the cost of an additional unit is different from the unit-by-unity costs you accumulated so far. Consider a drug company introducing a new drug to market: the cost to make the first tablet is enormous (the costs of researching, testing, and getting approval for the drug), while the cost to make the second tablet is much less: just leave the machine on for ten seconds longer. A similar example is the case of Amazon, who finds it cheaper to ship the millionth book than it did the first.[2] Conversely, consider asking a friend who works as a consultant for increasingly complex help: you might be able to get some brief advice for the cost of a cup of coffee, but if you want more than few minutes of thinking, you'll find that you'll start getting charged—indeed, more and more as the complexity of your problem becomes apparent. Another example is the declining marginal productivity of land: it's easier to feed the first fifty people in your village because you can farm the richest plots. As the population increases, you have to move to increasingly barren soil.[3]

For any marginal cost, you can ask: how much am I saving (or losing) compared to the cost I *would* have paid if that marginal cost applied at all levels? You can think of this as the foolish startup's price: if Amazon can find, pack, and ship a book for less than a dollar, my foolish friend reasons he can do the same with the books in his house. Or (if the curve goes the other way, with increments becoming more expensive) you can think of it as the freeloader's price; what happens when she notices that she only paid for an hour of advice, but actually got an hour and a half's worth once you count that free conversation over coffee. In the physical world, usually (but watch out!), it costs more to go from 50 to 51 miles/h than it does from zero to one—so we can think of this as a freeloader's price: I got up to 51 miles/h more cheaply on my Lagrange bill than I'd have expected given what I was charged for the last increment.

---

[2] Making this article completely self-referential, the first book shipped from Amazon was by Douglas Hofstadter and the Fluid Analogies group at Indiana.

[3] A more complicated relationship might obtain when hiring a taxi: the first mile is more expensive compared to the second mile, because you're usually charged a flat rate, or "flag drop". But if you try to get a taxi to take you from, say, Pittsburgh to Chicago, you'll find that you end up negotiating a much higher per-mile fee than you'd expect, since the driver won't be able to get a return fare. We'll focus here on the cases where every marginal cost specifies a unique unit amount—these also obtain for sensible physical laws, and our toy example of the drug firm. The technical term is that the Lagrangian is "non-degenerate".

Now, remarkably, the freeloader's gain—the difference between the freeloader's price and the actual price—has an interpretation in terms of the underlying physics. It's the energy!

Don't ask me *why* it's the energy. It does all the sorts of things we want energy to do, like total your car if you get too much of it too quickly. If you stick in a theory that you can solve some other way, where you've previously been able to identify the quantity you think is the energy, it always comes out the same. But why it should pop out of a crazy argument about linear extrapolations of marginal costs, I can't say in any simple, efficient way. You might as well say that there's a hidden economic structure to the Universe that we didn't expect, and it turns out that energy is just some quantity derived from that more fundamental structure.

In any case, and as long as the cost function doesn't change with time, you can prove that the total freeloader's gain in the system is constant—that's conservation of energy. Some people might get a few fewer hours of free consulting time, which lowers their freeloader's gain, but others, in turn, will get more. Something that we tend to think of as an essential quantity, neither created nor destroyed, ends up popping out of the Lagrangian formalism.

This story, about minimizing Lagrangian bills and freeloader's prices, may seem like an overly complicated way to talk about how particles move about in space. Famously, when the physicist Richard Feynman encountered it as an undergraduate during his physics education at MIT he rejected it entirely at first, coming up with increasingly ingenious ways to reason about physical systems using the standard set of Newton's laws. Why bother rephrasing the laws we already know in terms of a cost function? One answer is that it does provide a recipe for handling extremely complicated systems that you can't keep track of all at once. Even Feynman had, at some point, to switch over.

Another answer is that it provides a very general way to think about physical laws. All you need to do is specify that cost: a single equation, for example, can replace all three of Newton's laws of motion. Today, when inventing a new quantum field theory, all the physicist has to do is write down a Lagrangian. (Solving it, of course, is another problem altogether.) Most germane to our discussion, the Lagrangian formalism allows us to speculate on physical laws that have higher-derivative terms, providing a recipe book for how to interpret the equations in terms of physically real quantities like energy and momentum.

The effect of adding in these new terms is dramatic. This is because once you allow the Lagrangian to depend on more than just the velocity, but also the acceleration, you have multiple terms to consider. The Lagrangian depends on not just the cost of going from 50 miles/h to 51, but also the cost of going from (say) zero acceleration to 1 mile/h/s.

To continue the economic analogy, there are more goods to produce, and by adjusting the mixture of goods one produces, unexpected cost savings become possible. Following the standard recipes shows that it's now possible to find economies of scale: as one speeds up, for example, it becomes easier and easier to speed up more. We move from the village farm to the Amazon warehouse. These unexpected gains correspond to negative energies: rather than costing energy to get there, you actually release a little.

This sounds harmless at first. But there are still positive energy paths. And, although energy can neither be created nor destroyed, we're now in a system where particles can charge arbitrarily large gains to other particles, as long as they can match foolish startup prices to freeloader's gains. A car can accelerate to an arbitrarily high speed and high energy, from nothing, as long as it can find another car who can produce negative energy to keep the totals constant. You might say we've invented debt, and given the particles, like Lehman Brothers, no constraints on how much they can leverage.

5.5   A Conversation with John Bova, Dresden Craig, and Paul Livingston

The "Undecidables" [22] met on 5 July 2017 in Santa Fe to discuss a draft of this essay. John Bova, Dresden Craig, Paul Livingston and I participated, in a meeting that also touched on papers by John [23] and Simon Saunders [24]. In the following weeks, the group proposed a series of questions based on that conversation, which I've attempted to answer here.

Dresden Craig. In the abstract for this paper, you write that "the universe need not have made self-reference possible." Could you say a little more precisely what you mean by that? Do you mean "the universe" there to indicate a universe with the same fundamental laws of physics as our own, or are you also thinking about other possible universes? (If the latter, "possible" in what sense?) In either case, can we say anything meaningful about what a universe in which self-reference was impossible would be like?

SD. My main concern here was with universes that had laws, physical laws, that differed from ours but nonetheless could be expressed in the formalisms and mathematics we have to hand in our own. The story of modern physics is in part the story of how we've come to learn that—if we value mathematical coherence—this space is much more restricted than we used to think, and that's a lot of fun. There are certainly strange universes well beyond that, that philosophers have considered, but to include all of those as well is a bit cheating. It's trivial to consider, say, a universe consisting only of a perfect sphere, hanging unaided in an infinite space. There's not a lot going on there.

What you do need for self-reference are structures sufficiently densely-interlocked that they support an effectively unbounded memory. The natural way for this to happen, in our universe, is through the sticking together of stuff in increasingly complex ways, with memories being spread out over increasingly large distances. Consider the leap from a protocell, with a simple membrane to separate in from out, to the human brain. You don't even need a Lagrangian in your fundamental theory for that to happen: all sorts of crazy theories could do that, including whatever M-Theory turns out to be. But, conversely, it's possible to write down universes where this stuff would not happen: e.g., a billiard-ball universe without gravity.

It's worth noting that there are many ways to get this interlocking, some of which would seem very strange to us. Imagine, for example, things clustering together not in space, but in "momentum space"—nearby by virtue of having similar values of momentum, like cars grouped not by where they were on the highway, but by their speeds. You could imagine physical laws tuned in such a way that particles with particular momenta were able to preferentially interact even if widely separated (or delocalized) in space. Memory could emerge.

Paul Livingston. My first question is about physics, memory, and time. As you note in your paper, fundamental physics doesn't appear to allow for memory to be a basic (or even a real) phenomenon: since the time parameter in statements of physical law is always just given as a simple, single value increasing over time, there doesn't seem any warrant for introducing as physically real any operations of comparison between states of systems at distinct times. Things seem (at least at first) rather different from the perspective of computer science, where of course we constantly appeal to data being stored "in memory", and even Turing's basic architecture essentially includes an (ideally infinitely extended) symbolic memory.

At this level, we have memory in the sense of the ability to store syntactic symbols, and for the machine's functioning at one time to depend on what has previously been stored; but we don't yet seem to have a basis for at least some of the further emergent phenomena you discuss in your paper—for instance meaningful experience—until these symbols and comparisons are in fact "interpreted" by some kind of conscious subject or agent. If this is right, it would seem to make the presence of such an agent (who lives in experienced time) essential for these phenomena themselves, as if in an important sense it is us who are constituting or making up time (beyond just the single, linear time-parameter of basic physics).

In the history of philosophy, there's a long legacy of arguments that say that time is not basically real, or is illusory at the basic level of reality and is rather constituted by us as human subjects. These arguments perhaps begin with Zeno, who held, for example, that an arrow in flight cannot really be moving, since at each discrete moment (each discrete value of t) it is at rest. Others such as Kant and Husserl have seen time and the meaningful phenomena of change, motion, and causality as imposed by the form of our minds or our understanding upon the world, while still others (such as Bergson) have tried to re-introduce "memory" into matter by thinking of universal time, including physical time, as constituted in part by a kind of universal cosmic "evolution" toward progressively higher forms.

My question then is whether and how the dynamics of self-reference can allow us to see time and memory as really "there" (at the basic physical and/or computational levels) and not just constituted or produced by us. And can we maintain this kind of realism about time and memory without thinking that all of the progressive developments that you've invoked (life from non-life, mind from non-mind, and society from individuals) were "pre-inscribed," that is, already built in to the basic physics of the universe, somehow?

SD. Modulo some minor translations between our two languages, I'd agree with your account of time plays here, and your extension to the experience of conscious agents. It certainly feels like it's impossible to experience the passage of time without some kind of reference to past and future, and such comparisons become impossible for an agent without the memory to do so, or a subject without access to those memories.

Once we phrase it this way, I come down squarely on the side you attribute to Kant and Husserl. We can knock off Zeno right away, if with a cannonball from the 19th Century. The Danish physicist Hans Christian Ørsted showed that an electrical current—meaning the flow of electrons through a wire—could move the needle on a compass sitting a little distance away by inducing a magnetic field. We now believe that effect doesn't depend in any fundamental way on the inhomogeneity of the flow. If you pushed a perfectly uniform, charged rod, infinitely long, past the compass, a similar thing would happen, but now the set of discrete moments for that moving rod are identical to the case where it's stationary. The only thing you're doing to is altering the velocity, and yet that difference causes a needle to twitch. So velocity is real, since anything that plays a causal role has to be real, and since position is also real and velocity is the derivative of position with time, there is some non-trivial sense in which time is a physical thing that just exists, does something in the world. Position and velocity are two legs of a tripod, and the whole thing can only stand with time.

But at the same time, we do want to say that what *we* mean by time—the feeling of time flowing up towards a deadline, say, or away into the past after a parting from someone you love—is more than just a component of physical law. It essentially involves the awareness of change, internal or external, the ability to compare one moment to the next, and to consider its meaning. We can't understand the world, make sense of it, without grasping these more complex objects—objects that, by Ostrogradsky's theorem, are banned from being fundamental constituents of our world. They emerge. I'd be happy to claim Bergson as a fellow-traveller here, since I want to say that memory and the possibility of self-reference can emerge prior to a subject; they can become available, we can have them to hand.

Now, I can't quite parse the Subject from the computation. I wish I could. But we can split the physical and computational apart, and say that you need the computational bit to gain awareness of time, and that computational bit can't be "baked in" like the time that Ørsted discovered.

DC. You write that "the fundamental laws [of physics] are forgetful, referring to how third (or higher) time derivatives cannot correspond to physical realities; then, you write that "coarse graining gives the possibility of memory and later you "point to a crucial moment where they (i.e., each new level of complexity) all begin: not in the perception of detail, but in its selective destruction and lossy compression." Is this a fair paraphrase? Memory is made possible by a kind of coarse graining, and coarse graining is a throwing away of details; memory is, therefore, a particular form of forgetting. If so, then can we say the inability to derive certain key thresholds of complexity from fundamental physical laws is tantamount to an inability of physics to predict what about itself is forgettable? Can we generalize to say that any theory which works for a given level will be unable to predict which coarse grainings of itself will make sense?

SD. This is a lovely question—and a lovely suggestion. One has a feeling that (for example) humans are constantly surprised by the coarse grainings society places on them: categories of race and sex, class, and so forth. "I wouldn't have ever imagined you'd do that to us," you can hear people say at critical points in history, referring not to a particular other person, but rather to a system that they find themselves caught up in. The understanding we have of our own inner lives can't work out what society will do with collections of them.

It may well be the case that we recognize the emergence of new things only in retrospect, when we have the examples before us; and even then, only imperfectly. And of course we can disagree about which coarse-grainings are appropriate. Historians generate multiple accounts of the same events, which you can think of as incompatible coarse-grainings, and they find the clashes between these accounts to be sources of fertile discussion, rather than signs that something is seriously wrong with the project.

PM. Throughout your paper, I was interested in the way that issues about how we describe the phenomena theoretically interact with issues about how these phenomena actually are in themselves. For example, one can argue that a computer's memory register is only actually its memory as described from a certain (functional) perspective: from another perspective, it is just a physical configuration of matter. We might distinguish here between the first and second of Marr's levels of analysis for systems.[4] If we draw the distinctions this way, it seems that a system can only be "self-referential" (if it is) as described in a certain way: that is, for us to see it as self-referential we must describe it from a perspective that portrays some of its physical changes and quantities as "references" to things, and also draws some line around just that system as "itself."

This might seem to suggest that while the idea of self-reference can change the way we view some of these systems, it doesn't play a role in the basic, physical-level behavior of the system itself. Yet all of the phenomena you describe (life, mind,

---

[4]Marr, a neuroscientist, described three levels for the analysis of human visual processing: the "computational" (what purpose is the system achieving—recognizing faces, say), "algorithmic" (how does the system break that task into subtasks that fit together? can you write down the psuedocode of that process?), and "implementation" (how the brain actually gets things done, with Potassium ions and depolarization waves) [25].

and society, etc.) do certainly—once they are there—make a difference to the actual physical behavior of the relevant systems: for example, given a functioning society, matter will be moved to places it would not be if there were just the individuals acting without any conception of the larger whole. How should we understand, then, the actual reality of these features of self-reference and self-organization, at the most basic physical level? Or does acknowledging them require us to hold that a total explanation of the world written only in the vocabulary of the basic physical level (without terminology such as "reference"and "self-") could never capture all that goes on there?

SD. I think the Marr account is the correct one. We can get these terrific, efficient descriptions of how a pile of mail got from one side of the Atlantic to the other by referring to emergent properties. We could re-write everything in a more fundamental language, but why? We do so much better if we don't. You might say that we know the reality of reference and self from the epistemic resistance we encounter when we try to get rid of them.

John Bova. Is it possible to take us a little farther into the discussion of renormalization? And how do the considerations about renormalization relate to the simpler reasons that we might expect the fundamental laws of medium-sized nature (at least) to work on the order of second derivatives? For instance, it seems as though there ought to be a connection to how a conservative field can be understood as throwing away path-dependent information.

SD. There are many different ways things can become ignorable. When we write down a theory, we encounter quantities that are "truly" irrelevant: for example, in classical electrodynamics, the zero point of the electric potential. You can move those quantities around as much as you like, and the predictions of the model don't change. We call the gauge invariances. They were never there in the real world in the first place, but were rather ghosts born of our limitations. The structures in the world that we want to describe don't map perfectly onto the mathematical objects we know how to manipulate. There's an excess, though sometimes we find better notation that kills them off once and for all.

In other cases, we have facts about the world that are real, but causally irrelevant. The motion of a particle in a conservative field, as you note, can be both predicted and explained without reference to the entire path it took. It's not that the path doesn't have some kind of reality. You and I can watch the Space Shuttle launch, in a way that we can't, for example, observe the zero point of the electric potential. It's just that the path is irrelevant from the point of view of the physics of the phenomenon itself. A similar thing happens for jerk, in fundamental physics: it's not that you can't take the third derivative of position, it's just that it doesn't matter.

Renormalization is a third, distinct, way of ignoring things. In this case, you actually end up ignoring things that matter! You pick and choose carefully so that what you're ignoring is (for example) the least-damaging, least problematic stuff.

If you're interested in building a road, you don't need to have the positions of the quarks, but they're really there and they matter causally to what's going to happen to your road. Renormalization does a huge number of things in physics, beyond just the production of efficient descriptions and (as we use it in this paper) the discovery of higher-order, non-local interactions. One of the first reasons we created it was to handle some problems with a theory that we thought was fundamental (quantum electrodynamics, QED) but had all sorts of problems when we tried to treat it that way. In the end, we had to create the Higgs boson—but we were able to do QED calculations before we knew what the Higgs was, because we could treat QED as a theory that came out of some mysterious mystery theory deeper down.

I'd be remiss if I didn't mention the Santa Fe Institute MOOC we did on this [26]; one of the papers on renormalization in complex, computational systems that we talk about there comes from Israeli and Goldenfeld [27].

JB. Will a human sorter faced with the Shakespearean task produce a locally sharp but globally indefinable shape, or will they produce a hopelessly vague shape even at the level of local decisions? What does the answer tell us about how to apply theorems on formal systems to human intentional states and acts without conflating or merely analogizing the two? What does it tell us about where intentional states are located in conceptual space relative to consistency and completeness?

SD. You're asking a crucial question. Paul asks it from a slightly different angle, making a distinction between the computation and the subject, while you distinguish formal system from human reason. How do we, as subjects, differ from our computational states? Or (a more restricted question) where are *we* located, if we are at all, on the Marrian levels? The purpose of this essay is to give a story about how a certain set of objects (memory, self-reference) that are necessary (but not sufficient) for meaning, come to be, and it's tempting to punt the rest to biology, tying the intentionality and meaning of human states to evolution. But let's try a bit harder before we punt.

What fails when we think that formal systems are identical to human states of mind, thoughts a subject thinks? An obvious answer is the latter are subject to continuous revision. I may hold in mind the same object, person, thought, at two different times, and have them mean, or imply, completely different things. We develop, over time, in ways that I think can't be simply reduced down to computational talk about changing variables, or even LISP-like metaprogramming stories. We're not substituting, we're returning.

Let answer this synthesis of your question and Paul's with an answer suggested by Dresden. What if we take seriously the revisability of our our mind's meanings, the ways in which we can return to the same states, the same thoughts, and have them mean completely different things to us? Here let me include our emotional responses, our feelings, as well as our more conceptual, intellectual states.

We could say, well, it's just the valence that's changed, the thing is the same but it's stuck in a different place of our affective network. The context is different. But I think it's more than this: there's something unsatisfying about postulating a world of atomic thoughts that we combine together like Lego bricks, even if we made those bricks ourselves. At the very least, we experience a shift in emotional gestalt, where we might in some way be able to say the object is identical, but not in any way that actually matters. We feel love for a person, but where this was originally a pleasure, it has now become a pain, because that person is gone from us in some way. And it becomes a pleasure or a pain in a very different way from how we perceive food differently depending on how hungry we are.

Put these pieces together and we have a problem. How can we have a simple object (love, say—but if you like, one of the components of love) that persists in time (the love now is the love then), but also changes in some essential quantities (it was a pleasure, now it's bittersweet)? One answer is if the coarse-graining of the fundamental constituents is changing. At the level you're experiencing love, it's a simple object, whose complexity is hidden from you by the coarse-graining process. Events shock you, knock you about, and that coarse-graining has shifted in scale. The nature of that shift allows you to track the object over time, as its properties change, or at least to rediscover it after a shock. Perhaps it zooms out, so that the difficult parts of that love become invisible; perhaps it zooms in a bit, magnifying feelings you didn't know you had; most likely, some combination of the two. The shapes of our minds, and how they map to some computational or formal-language account, are not just vague. They're shifting. That makes formal systems inadequate not just as descriptions of our actual function (we already knew we weren't perfect reasoners), but also as normative accounts of our emotional lives.

References

1. Turkle, S.: Life on the Screen. Simon and Schuster, New York, NY, USA (2011)
2. Walker, S.I., Davies, P.C.W.: The algorithmic origins of life. J. R. Soc. Interface **10**(79), 20120869 (2013)
3. Kauffman, S.A.: Humanity in a Creative Universe. Oxford University Press, Oxford, United Kingdom (2016)
4. Hobson, E.A., DeDeo, S.: Social feedback and the emergence of rank in animal society. PLoS Comput. Biol. **11**(9), e1004411 (2015)
5. DeDeo, S.: Major transitions in political order. In: Walker, S.I., Davies, P.C.W., Ellis, G.F.R. (eds.) From Matter to Life: Information and Causality. Cambridge University Press, Cambridge, United Kingdom (2017). https://arxiv.org/abs/1512.03419
6. Aaronson, S., Carroll, S.M., Ouellette, L.: Quantifying the rise and fall of complexity in closed systems: the coffee automaton. arXiv preprint cond-mat.stat-mech (2014). https://arxiv.org/abs/1405.6903
7. Scott, A.: Why philosophers should care about computational complexity. In: Computability: Turing, Gödel, Church and Beyond, pp. 261–328. MIT Press, Cambridge, MA, USA (2013)
8. Chomsky, N.: Aspects of the Theory of Syntax. MIT Press, Cambridge, MA, USA (2014). (1964)
9. Smith, J.M., Száthmary, E.: The Major Transitions in Evolution. Oxford University Press (1997)
10. Fukuyama, F.: The Origins of Political Order: From Prehuman Times to the French Revolution. Farrar, Straus and Giroux (2011)

11. Smail, D.L.: On Deep History and the Brain. University of California Press (2007)
12. Donald, M.: An Evolutionary Approach to Culture. In: Bellah, R.N., Joas, H. (eds.) The Axial Age and its Consequences. Harvard University Press (2012)
13. Graeber, D.: Debt: the First 5,000 Years. Melville House, New York, NY, USA (2014). (Updated and expanded)
14. Woodard, R.P.: The theorem of Ostrogradsky (2015). arXiv:1506.02210. https://arxiv.org/abs/1506.02210
15. DeDeo, S., Psaltis, D.: Stable, accelerating universes in modified-gravity theories. Phys. Rev. D **78**(6), 064013 (2008)
16. Cooney, A., DeDeo, S., Psaltis, D.: Gravity with perturbative constraints: dark energy without new degrees of freedom. Phys. Rev. D **79**(4), 044033 (2009)
17. DeDeo, S.: Effective theories for circuits and automata. Chaos Interdiscip. J. Nonlinear Sci. **21**(3), 037106 (2011)
18. Flack, J.C.: Multiple time-scales and the developmental dynamics of social systems. Philos. Trans. R. Soc. B Biol. Sci. **367**(1597), 1802–1810 (2012)
19. Flack, J.C.: Coarse-graining as a downward causation mechanism. Philos. Trans. R. Soc. Lond. A Math. Phys. Eng. Sci. **375** (2109) (2017)
20. Munroe, R.: Nerd-swiping. http://xkcd.com/356
21. Woodard, R.P.: Ostrogradsky's theorem on hamiltonian instability. Scholarpedia **10**(8), 32243 (2015). (revision #151184)
22. Bova, J., Craig, D., DeDeo, S., Livingston, P.: The Undecidables (2013). http://tuvalu.santafe.edu/~simon/undecidables.txt
23. Bova, J.: Groups as eide? toward a Platonic response to Metaphysics M on unity, structure, and number (2017). (Unpublished manuscript)
24. Saunders, S.: Physics and leibniz's principles. In: Brading, K., Castellani, E. (eds.) Symmetries in Physics: Philosophical Reflections. Cambridge University Press, Cambridge, United Kingdom (2003)
25. Marr, D.: A Computational Investigation into the Human Representation and Processing of Visual Information. Freeman and Company, San Francisco, CA, USA (1982)
26. DeDeo, S.: Introduction to renormalization. Complexity Explorer MOOC, Santa Fe Institute (2017). http://renorm.complexityexplorer.org
27. Israeli, N., Goldenfeld, N.: Coarse-graining of cellular automata, emergence, and the predictability of complex systems. Phys. Rev. E **73**, 026203 (2006)

Chapter 6
Agent Above, Atom Below: How Agents Causally Emerge from Their Underlying Microphysics

Erik P. Hoel

6.1 Agents Excluded

Marco Polo describes a bridge, stone by stone.

'But which is the stone that supports the bridge?' Kublai Khan asks.

'The bridge is not supported by one stone or another,' Marco answers, 'but by the line of the arch that they form.'

Kublai Khan remains silent, reflecting. Then he adds: 'Why do you speak to me of the stones? It is only the arch that matters to me.'

Polo answers: 'Without stones there is no arch.

— Italo Calvino [1]

Agents form part of the ontology of our lives. From our daily interactions with entities that evince goals, intentions, and purpose, we reasonably conclude that agents both exist and act as a causal force in the world. And yet there is a dissonance between our easy belief in the ubiquity of agents and what we know of physics. Down at the level of the Hamiltonian, the level of the quark, all of reality is describable as merely lawful changes in the state of incredibly small components. Even if these many-body systems are hopelessly unpredictable in practice, or exist in unique or interesting configurations, all the individual components still follow strict and purposeless laws concerning their evolution in time. Down at this level we see no sign of agents, their intentions, goal-oriented behavior, or their evolved pairing with their environments.

E. P. Hoel (✉)

Department of Biological Sciences, Columbia University, New York, NY, USA

e-mail: hoelerik@gmail.com

© Springer International Publishing AG, part of Springer Nature 2018

A. Aguirre et al. (eds.), *Wandering Towards a Goal*, The Frontiers Collection,

https://doi.org/10.1007/978-3-319-75726-1_6

I'd like to begin by arguing that this dissonance between purposeful agent-like behavior at the top, and purposeless and automaton-like behavior at the bottom, speaks to a larger issue. While many systems lend themselves to descriptions at higher scales, there is always a base scale that underlies, or fixes, all those above it. Given just the workings of the base, one could, at least in theory, derive all scales above it.

In the language of analytic philosophy this relationship is called *supervenience*: given the lower-level properties of a system, the higher-level properties necessarily follow [2]. The base scale is the microphysical scale (the finest possible representation of the system in space and time). Deriving the scales above it can be conceptualized as mappings from the base microscale to supervening macroscales, where each mapping is over some set of states, elements, functions, mechanisms, or laws. Broadly, a macroscale is any scale that contains fewer elements or states, or simpler laws or functions. A well-known example of a macroscale is a coarse-grain, such as temperature versus the individual motion of particles. That is, the astronomical set of possible microstates constituting all combinations of particle kinetic energy can be coarse-grained into the single macrostate of temperature. In this case the mapping from the states of the individual particles to the temperature of the whole is a function (the average) of the kinetic energy. Not all macroscales are coarse-grains; they might be a subset of the state-space, like a black box with inputs and outputs (see Appendix A.1 for technical details). One universal way to think of macroscales is to consider the runtime of a full simulation of the system: moving up in scale decreases runtime.

Explicitly considering these relationships across scales reveals a fundamental issue. Given that supervening macroscales can all be derived from the base microscale, it would seem natural that some form of Occam's razor applies. That is, the higher-scale descriptions aren't necessary. The most aggressive form of the argument is known as the 'exclusion argument.' Given that all systems are describable at the physical microscale, what possible extra causal work is there for any supervening macroscale [3]? And if those higher scales don't contribute causal work then they are by definition epiphenomenal: shadows of what's really going on. So it seems that the universe collapses like a house of cards down to just the purposeless events and causal relationships of the physical microscale.

While this may seem a rather abstract concern, I think it can be framed as a triad of issues not confined to philosophical ruminations: model choice (microscale models are always better, at least in principle), causation (all causal influence or work is at the base microscale, at least in principle), and also information (no new information can be gained by moving up in scale, at least in principle).

In this broadened scope the issue of whether reduction is always in principle justified affects all of science, especially the so-called "special sciences," like biology and psychology, which seem to comfortably float above physics as their own distinct domains. The special sciences are a reflection of one of the remarkable but often unremarked upon aspects of science: its large-scale structure.

Science forms a kind of ladder of ascending spatiotemporal scales (or levels), wherein each scale supervenes on those below it (Fig. 6.1). Yet even in this hierarchical structure all the information and causal work seems to drain away down to the

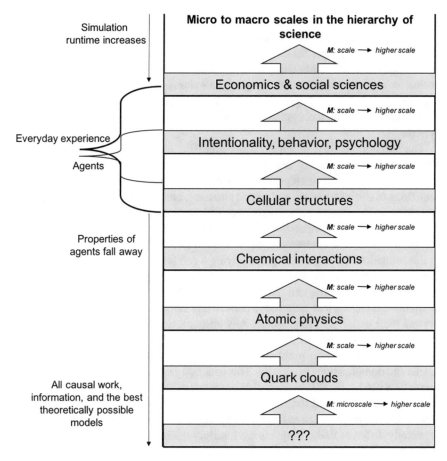

Fig. 6.1 The hierarchy of sciences arranged via mapping functions from one science to the next. Sciences like physics and chemistry are closer to the bottom, while sociology and psychology occupy the top

microscale [4, 5]. Not only that, but the exclusion of higher scales don't even require a final base microscale of physics. Rather, each rung could be causally reducible to the rung below it.

Agents are somewhere above biological mechanisms but below economics on the ladder. They are a major part of the slim section that corresponds to the scale of our everyday experiences. Agents themselves can be described, both mathematically and not, in agent-based modeling [6], in game theory [7], in behavioral analysis [8], biology [9], economics [10], and many other fields. While one might use detailed mechanistic explanations to construct the mapping function from the scale of quarks up to the scale of agent states or behaviors, this would beg the more perplexing question: how can something be both an agent at one level and not an agent at another? How can agents *really* exist when they are always definable in terms of goal-free and purposeless dynamics?

After all, as long as supervenience holds then agents appear to be just one more reducible higher-level representation of the underlying physics. At some point down the ladder all agent properties disappear. So while no one questions that some physical systems can be described in terms of their goals, intentional states, and behavior, the exclusion argument means that the purposelessness of microscale physics reigns supreme. And it is these purposeless dynamics which theoretically make up the most informative model of the system and which are doing all the causal work in the system.

The reductionist, one might even say nihilistic, answer to this conceptual knot can be stated clearly using the terminology of information theory: compression. Macroscales are useful or elegant summarizations. They are, at best, lossless representations of the information and causal structure, but are far more likely to be lossy and thus missing crucial information or causal relationships. Their usefulness stems from the necessity of compression in communication, because all systems have a limited bandwidth. Macroscales make the world summarizable and thus understandable. In this reductionist view, a biologist studying a cell is really referring to some astronomically complex constellation of quarks. It is merely because of the limited data storage, cognitive ability, and transmission capabilities of the human brain and the institutions it creates that we naively believe that cells exist as something above and beyond their underlying microscale. Any abstractions or higher-level explanations assist our understanding only because our understanding is so limited.

This formulation of the problem in terms of higher and lower scales puts aside the details and complications of physics. This generality is important because agenthood seems likely to be a multiply-realizable property: different microscopic scales, laws, or mechanisms may lead to the same agent behavior. For instance, artificial agents created in simulations, such as Conway's the Game of Life [11], may have underlying drastically different rules. If agents are multiply realizable in this manner, then any attempts to link the unique properties of agents to some unique property of our physics is doomed to underdetermination.

Recently, when asked, "What concept should be better known in science?" Max Tegmark answered: a form of multiple-realizability called substrate-independence. He gave the example of waves propagating in different mediums, and remarked that "we physicists can study the equations they obey without even needing to know what substance they are waves in" [12].

This property of multiple-realizability at first seems itself a good enough reason to not immediately exclude macroscales as mere epiphenomena [13]. But it's more suggestive than it is definitive. Unless, that is, the fact that something is multiply realizable means it is somehow doing extra causal work, or somehow contains more information? Recent research has argued exactly this [14, 15] by demonstrating the possibility of *causal emergence*: when a macroscale contains more information and does more causal work than its underlying microscale.

I'll first introduce causal emergence and then argue that agents not only causally emerge, but that significant aspects of their causal structure cannot be captured by any microphysical model.

6.2 Causal Emergence

> We may regard the present state of the universe as the effect of its past and the cause of its future.—Pierre-Simon Laplace [16]

Demonstrating causal emergence requires causally analyzing systems across a multitude of scales. Luckily, causal analysis has recently gone through an efflorescence. There is now a causal calculus, primarily developed by Judea Pearl, built around interventions represented as an operator, $do(X = x)$ [17]. The operator sets a system into a particular state (at some time t) and then one can observe the effect at some time t_{+1}. This can be used to assess the causal structure of systems at different scales by applying either a macro or micro intervention. For instance, a micro intervention might be setting the individual kinetic energy of all the particles in a gas, while a macro intervention would be fixing the average kinetic energy (temperature) but not specifying what the individual particles are doing (see Appendix A.1 for further details).

For the following proof-of-principle argument, let's just assume that all systems we're talking about are discrete, finite, and have a defined microscopic scale. To understand the causal structure of such a system, a set of interventions is applied to reveal the effect of each state. This is represented as an intervention distribution (I_D) of individual $do(X = x)$ operators, which will also generate a set of effects (future states), the effect distribution (E_D). The causal structure of a system can be inferred from the application of I_D (and a liberal helping of Bayes' rule) to derive the conditional probabilities $p(x|y)$ for all state transitions. Notably, the conditional probabilities can change significantly at the macroscale. For example, consider the transition matrix of a toy system (or causal model) with a microscale describable as a Markov chain:

$$S_{microscale} = \begin{bmatrix} 1/3 & 1/3 & 1/3 & 0 \\ 1/3 & 1/3 & 1/3 & 0 \\ 1/3 & 1/3 & 1/3 & 0 \\ 0 & 0 & 0 & 1 \end{bmatrix} \xrightarrow{yields} S_{macroscale} = \begin{bmatrix} 1 & 0 \\ 0 & 1 \end{bmatrix}$$

where a macroscale is constructed of a grouping (coarse-grain) of the first three microstates (it is multiply realizable). Notice how the conditional probabilities have changed at the macroscale: transitions are more deterministic and also less degenerate (less mutual overlap in transitions). This means that interventions at the macroscale are more sufficient in producing a specific effect and more necessary for those effects [14].

Quantifying this micro-to-macro change can be done using information theory. This is possible because causal structure is a matrix that transforms past states into future states; it thus can be conceived of as an information channel. In this view, individual states are like messages and interventions are like the act of sending them over the channel [15]. We can then ask how much each state (or intervention) reduces the uncertainty about the future of the system by measuring the mutual information between some intervention distribution and its resultant effect distribution: $I(I_D;E_D)$. It's an equation that quantifies Gregory Bateson's influential definition of information as "a difference that makes a difference" [18]. The method of intervening on system states and assessing the mutual information between past and future states has been called *effective information* [14, 19] and reflects the determinism and degeneracy of the system's causal structure (see Appendix A.2). In the above toy case, over the full set of possible interventions (all possible states) the microscale has 0.81 bits, while the macroscale has 1 bit.

Where does this extra information come from? If we consider the original Markov chain above it's obviously not a perfect information channel. Rather there is noise and overlap in the state transitions. This might be irreducible to the channel, or it might be a function of the measurement device, or the channel might be an open system and the outside world provides the noise. Notably, in 1948 Claude Shannon showed that even noisy channels could be used to reliably transmit information via channel coding [20]. His discovery was that channels have a particular capacity to transmit information, which in noisy channels can only be approached by using codes that correct for errors (such as using only a subset of the messages): $C = max_{p(X)}I(X;Y)$.

I've argued there is an analogous causal capacity of systems to transform sets of states (such as input states, internal states, or interventions) into future states. Continuing the analogy, the input $p(X)$ is actually some intervention distribution $p(I_D)$. Under this view different scales operate exactly like channel codes, and the creation of macroscales is actually a form of encoding (Appendix A.3). Furthermore, models at different scales capture the causal capacity of a system to different degrees. This means we can finally specify exactly what makes multiply-realizable entities (sometimes) do more causal work and generate more information: they provide error-correction for causal relationships and thus can be more informative than their underlying microscale, a phenomenon grounded in Shannon's discovery of error-correcting codes [15].

If this is true then causal emergence may explain why science has the hierarchal large-structure that it does. New rungs in the ladder of science causally emerge from those below them. Different scientific fields are literally encodings by which we improve our understanding of nature.

6.3   Agents and the Scale of Identity

If it isn't literally true that my wanting is causally responsible for my reaching and my itching is causally responsible for my scratching... if none of that is literally true, then practically everything I believe about anything is false and it's the end of the world.

— Jerry Fodor [21]

Many types of systems can demonstrate causal emergence, meaning that the full amount of information and causal work can only be captured at a macroscale. Some of these macroscales can be classified as agents that operate via purposes, goal-oriented behavior, intentional states, and exist in intelligent or paired relationships to their environment. It is actually these unique agential properties that prime agents to causally emerge from their underlying microscales.

For instance, agents are stable at higher spatiotemporal scales but not lower ones. They maintain their identity over time while continuously changing out their basic constituents. While the exact numbers and timescales are unknown, nearly every component part of you, from atoms to cellular proteins, are replaced over time. Like Theseus' ship, agents, such as Von Neumann machines or biological life, show a form of self-repair and self-replacement that is always ongoing, particularly at lower scales. This self-maintenance has been called autopoiesis and is thought to be one of the defining aspects of life and agents [22], a kind of homeostasis but for maintaining identity. As autopoietic systems, agents need to intake from the world for metabolic reasons, thus locally increasing the entropy around them [23]. Most importantly, during this process the supervening macroscale of an agent is much more likely to remain stable than its underlying physical microscale, precisely because it is multiply realizable. This is particularly true of the macroscale causal relationships, which may be stable even as their underlying microscale causal relationships are ephemeral.

Consider the simple system shown in Fig. 6.2, where it is clear that only the supervening macroscale stays causally relevant over time because of autopoiesis. Rather than specifying the specific micro-element identities (A, B, C…) at each new timestep of the microscale, we can just specify the stable macro-elements (α, β, γ…) via some relational mapping. Defining the system in terms of its relations (rather than the identities of individual elements) seems to fit well the definition of a macroscale: if we again think about scale in terms of simulation time, relational descriptions don't need to be constantly updated as new components replace the identities of old components; thus their runtime is shorter than the microscale.

Fine, a fervent reductionist might answer, biting the bullet. While we must continuously update our microscale causal model of the system, there is still no information or causal work to be gained.

Of course, we already know this can be untrue because of causal emergence. However, there are even cases of causal emergence where there is a causal relationship at a relationally-defined macroscale without any underlying microscale. The macro-elements still supervene on the micro-elements, but the macroscale causal relationships don't directly supervene on any underlying microscale causal relationships.

A thought experiment makes the point. Imagine an agent that's quite large (many microscopic components) and these components have a high replacement rate r. Like some sort of amoeba, it internally communicates by waves spreading across its body; what matters in this contrivance is that different parts of its body might have a long lag time in their causal relationships. Now imagine that the lag time (the propagation timescale t) is actually greater than the average turnover rate of the microscopic building blocks. At the microscale where identity is always being updated (as A is replaced by Z, and so on), the causal relationships for which $t > r$ wouldn't even exist, as the receiver would always drop out of the agent before the intervention on

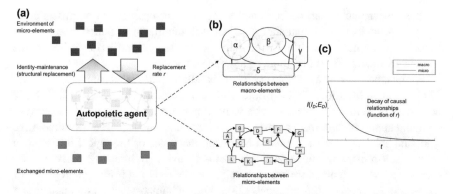

Fig. 6.2 *A toy autopoietic agent.* **a** The agent engages in 'feeding' behavior in which it replaces its micro-elements with those from the surrounding environment. **b** The supervening macroscale causal relationships, compared to those of the underlying microscale. **c** Pairs of causal relationships either rapidly decay at the microscale as elements enter and leave the agent (after exiting, an intervention on one doesn't affect the other). However, they are stable at the macroscale. The response to sets of interventions would correspondingly decrease their informativeness over time, but only at the microscale

the sender could reach it. To give a suggestive example in plain English, a description like "intervening on the far left element triggers a response in the far right element" wouldn't be possible in the less-abstract microscale language of "intervening on element A always triggers a response in Z" because Z is always replaced before it can respond.

In general, we can refer to the reductive instinct that ignores this type of problem as *causal overfitting*: when a representation restricted to the least-abstract base microscale fails to capture the causal structure appropriately. This overfitting can even completely miss causal work and information entirely: there may be causal relationships that exist solely at the level of agents. It points us to a good (if cloyingly tautological) definition of when something that seems like an agent *really* is an agent: it's when the only way of avoiding causal overfitting it to consider the system in terms of homeostatic, representational, or intentional states.

6.4 Teleology as Breaks in the Internal Causal Chain

> Romeo wants Juliet as the filings want the magnet; and if no obstacles intervene he moves towards her by as straight a line as they. But Romeo and Juliet, if a wall be built between them, do not remain idiotically pressing their faces against its opposite sides like the magnet and the filings... Romeo soon finds a circuitous way, by scaling the wall or otherwise, of touching Juliet's lips directly. With the filings the path is fixed; whether it reaches the end depends on accidents. With the lover it is the end which is fixed, the path may be modified indefinitely.
>
> — William James [24]

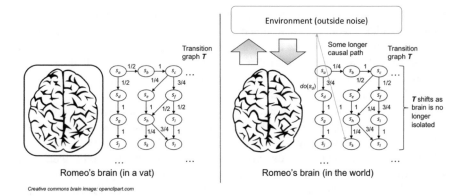

Fig. 6.3 *Teleological causation*. (**Left**) Romeo's brain and its transition graph while separated from the environment. There is no possible path of any length from s_d to s_k. (**Right**) Shown in red is a longer transition which eventually connects s_d to s_k but only operates because the agent is an open system. It appears merely as if the noise transitions have changed (over some further down/future transitions) but really a deterministic causal relationship has been outsourced to the environment

The purposeful actions of agents are one of their defining characteristics, but are these intentions and goals actually causally relevant or are they just carried along by the causal work of the microscale? As William James is hinting at, the relationships between intentions and goals seem to have a unique property: their paths can be modified indefinitely. Following the logic above, they are causally relevant because as causal relationships they provide for error-correction *above and beyond* their underlying microscales. How does this feed into the real, or perhaps merely apparent, teleology of agents?

One of the defining aspects of agents is that they are open systems. For example, the most reductive model of an agent would be a self-contained microphysical system, i.e., some set of microphysical states along with the associated possible state-space trajectories. But reductively restricting our view to just the internal structure of an agent can lead to *causal underfitting*. This in turn gives us a good definition of teleological causation.

Consider the case of Romeo's brain. While keeping in mind this is a drastic simplification, we can still represent it as some finite set of internal states, specifically as a Markov process: a finite stochastic model made up of states $\{s_a, s_b, s_c, \ldots\}$ and governed by some transition matrix T, so it can be represented as a transition graph (Fig. 6.3).

If Romeo's brain is cut off from its environment, let's also assume that as a Markov process it is reducible, so it's not always possible to get from one state to any other state. So if there is a state s_d, which is Romeo's desire to kiss, and a state s_k, which is kissing (or the feeling of a kiss), there may be no path between them. Which further means that some experimenter would conclude there's no possible causal relationship between s_d and s_k in Romeo's brain. So imagine the surprise of the same experimenter intervening to set Romeo's brain into s_d in a situation where Romeo is no longer isolated from his environment (and his Juliet). Struck by Cupid's arrow, Romeo will indefinitely pursue his goal of kissing Juliet, and to the experimenter's surprise s_d will inexorably, almost magically, always lead to s_k. It will appear as if teleology stepped in and ushered the system along a deterministic causal path that didn't exist given just the properties of the system itself.

Considering just the internal architecture of Romeo's brain gives no hint of all the causal relationships that actually exist within it, because it has outsourced so much of its causal structure to the environment. Note that for these causal paths to be reliable despite the vagaries of the environment they must be macro. Furthermore, we can identity what constitutes a teleological causal relationship: it's when causal relationships don't supervene locally on the system. The causal structure simply doesn't inhere to the system itself; it is only when the state of the entire environment is taken into account does it appear. Regardless of whether this is truly teleology, or rather just the local appearance of it, the source is the purposeful and goal-oriented behavior of Romeo as an agent.

6.5   Agents Retained

The fundamental problem of communication is that of reproducing at one point either exactly or approximately a message selected at another point.

— Claude Shannon [20]

For any given system, the world is a noisy place (see Appendix A.4). If the system's causal structure is constructed down at the level of tokens, rather than encoded into types, it will be unpredictably buffeted by the slightest outside perturbation. As we've seen, changes in scale are a form of channel coding, and by encoding up at macroscales causal relationships can error-correct, do more work, generate more information, and offer a better model choice for experimenters. It may be that scientists *en masse* implicitly choose certain scales for this reason, and the creation of a hierarchy of science is truly following Plato's suggestion to "carve nature at its joints."

Agents are exemplars of causal emergence; they are codes in nature that use to their advantage being open and autopoietic and having goal-oriented behavior, thus allowing for stable-over-time and teleological causal relationships. Ultimately, this means that attempting to describe an agent down at the level of atoms will always be a failure of causal model fitting. Agents really do exist and function, despite admitting of goal-free and purposeless microscale descriptions. While these descriptions are technically true, in the framework developed here the purposeless microscale descriptions are like a low dimensional slice of a high dimensional object (Appendix A.5).

Acknowledgements I thank Giulio Tononi, Larissa Albantakis, and William Marshall for our collaboration during my PhD. The original research demonstrating causal emergence was possible [14] was supported by Defense Advanced Research Planning Agency (DARPA) Grant HR 0011-10-C-0052 and the Paul G. Allen Family Foundation.

Appendix A. Technical Endnotes

A.1 Scales and Interventions

To simplify, only discrete systems with a finite number of states and/or elements are considered in all technical endnotes. The base microscopic scale of such a system is denoted S_m, which via supervenience fixes a set of possible macroscales $\{S\}$ where each macroscale is some S_M. This is structured by some set of functions (or mappings) $M : S_m \rightarrow S_M$ which can be over microstates in space, time, or both.

These mappings often take the form of partitioning S_m into equivalence classes. Some such macroscales are coarse-grains: all macrostates are projected onto by one or more microstates [14]. Other macroscales are "black boxes" [25]: some microstates don't project onto the macrostate so only a subset of the state-space is represented [15, 26]. Endogenous elements at the microscale (those not projected onto the macroscale) can either be frozen (fixed in state during causal analysis) or allowed to vary freely.

To causally analyze at different scales requires separating micro-interventions from macro-interventions. A micro-intervention sets S_m into a particular microstate, $do(S_m = s_m)$. A macro-intervention sets S_M instead: $do(S_M = s_M)$. If the macrostate is multiply-realizable then a macro-intervention corresponds to:

$$do\left(S_M = s_M\right) = \frac{1}{n} \sum_{s_{m,i} \in s_M} do\left(S_m = s_{m,i}\right)$$

where n is the number of microstates (s_i) mapped into S_M.

A.2 Effective Information and Causal Properties

Effective information (*EI*) measures the result of applying some *Intervention Distribution* (I_D), itself comprised of probabilities $p(do(s_i))$ which each set some system S into a particular state s_i at some time t. Applying I_D leads to some probability distribution of effects (E_D) over all states in S. For systems with the Markov property each member of I_D is applied at t and E_D is the distribution of states transitioned into at t_{+1}. For such a system S then *EI* over all states is:

$$EI\,(S) = \frac{1}{n} \sum_{s_i \in S} D_{KL}\,(S_F\,|do\,(S = s_i)\,\|E_D\,)$$

where n is the number of system states, D_{KL} is the Kullback-Leibler divergence [27], and $(S_F\,|do(S = s_i))$ is the transition probability distribution at t_{+1} given $do(S = s_i)$. Notably, if we are considering the system at the microscale S_m, EI would be calculated by applying I_D uniformly (H_{max}, maximum entropy), which means intervening with equal probability ($p(do(s_i)) = 1/n$) by setting S into all n possible initial microstates ($do\,(S = s_i)\,\forall_i \in 1 \ldots n$). However, at a macroscale I_D may not be a uniform distribution over microstates, as some microstates may be left out of the I_D (in the case of black boxing) or grouped together into a macrostate (coarse-graining).

Notably, EI reflects important causal properties. The first is the determinism of the transition matrix, or how reliable the state-transitions are, which for each state (or intervention) is:

$$D_{KL}\,(S_F\,|do\,(S = s_i)\,\|H_{max}\,)$$

While the degeneracy of the entire set of states (or interventions) is: $D_{KL}\,(E_D\|H_{max})$. Both determinism and degeneracy are [0, 1] values, and if one takes the average determinism, the degeneracy, and the size of the state-space, then: $EI = (determinism - degeneracy) * size$.

A.3 Scales as Codes

The capacity of an information channel is: $C = max_{p(X)} I\,(X; Y)$, where $I(X;Y)$ is the mutual information $H(X) - H(X|Y)$ and $p(X)$ is some probability distribution over the inputs (X). Shannon recognized that the encoding of information for transmission over the channel could change $p(X)$: therefore, some codes used the capacity of the information channel to a greater degree.

According to the theory of causal emergence there is an analogous causal capacity for any given system: $CC = max_{(I_D)} EI\,(S)$.

Notably, for the microscale S_m $I_D = H_{max}$ (each member of I_D has probability $1/n$ where n is the number of microstates). However, a mapping M (described in Appendix A) changes I_D (Appendix B) so that it is no longer flat. This means that EI can actually be higher at the macroscale than at the microscale, for the same reason that the mutual information $I(X;Y)$ can be higher after encoding. Searching across all possible scales leads to EI_{max}, which reflects the full causal capacity. EI can be higher from both coarse-graining [14] and black-boxing [15].

A.4 What Noise?

If the theory of causal emergence is based on thinking of systems as noisy information channels, one objection is that real systems aren't actually noisy. First it's worth noting that causal emergence can occur in deterministic systems that are degenerate [14]. Second, *in practice* nearly all systems in nature are noisy due to things like Brownian motion. Third, any open system receives some noise from the environment, like a cell bombarded by cosmic rays. If one can only eliminate noise by refusing to take any system as *that* system, this eliminates noise but at the price of eliminating all notions of boundaries or individuation. Fourth, how to derive a physical causal microscale is an ongoing research program [28], as is physics itself. However, it is worth noting that *if* the causal structure of the microscale of physics is entirely time-reversible, *and* the entire universe is taken as a single closed system, then it is provable that causal emergence for the universe as a whole is impossible. However, as Judea Pearl has pointed out, if the universe is taken as a single closed system then causal analysis itself breaks down, for there is no way to intervene on the system from outside of it [17]. Therefore, causal emergence is in good company with causation itself in this regard.

A.5 Top-Down Causation, Supersedence, or Layering?

To address similar issues, others have argued for *top-down causation*, which takes the form of contextual effects (like wheels rolling downhill [29]), or how groups of entities can have different properties than individuals (water is wet but individual H_2O molecules aren't). Others have argued that causation has four different Aristotelian aspects and different scales fulfill the different aspects [30]. It's also been suggested that the setting of initial states or boundary conditions constitute evidence for top-down causation [31], although one might question this because those initial states or boundary conditions can themselves also be described at the microscale.

Comparatively, the theory of causal emergence has so far been relatively metaphysically neutral. Openly, its goal is to be intellectually useful first and metaphysical second. However, one ontological possibility is that causal emergence means the macroscale supersedes (or overrides) the causal work of the microscale, as argued originally in [14]. A different metaphysical option is that scales can be arranged like a layer cake, with different scales contributing more or less causal work (the amount irreducible to the scales below). Under this view, the true causal structure of physical systems is high dimensional and different scales are mere low dimensional slices.

References

1. Calvino, I.: Invisible cities. Houghton Mifflin Harcourt (1978)
2. Davidson, D.: Mental events. Reprinted in Essays on Actions and Events, 1980, 207–227 (1970)
3. Kim, J.: Mind in a physical world: an essay on the mind-body problem and mental causation. MIT Press (2000)
4. Bontly, T.D.: The supervenience argument generalizes. Philos. Stud. **109**(1), 75–96 (2002)
5. Block, N.: Do causal powers drain away? Philos. Phenomenol. Res. **67**(1), 133–150 (2003)
6. Castiglione, F.: Agent based modeling. Scholarpedia **1**(10), 1562 (2006)
7. Adami, C., Schossau, J., Hintze, A.: Evolutionary game theory using agent-based methods. Phys. life Rev. **19**, 1–26 (2016)
8. Skinner, B.F.: The Behavior of Organisms: An Experimental Analysis (1938)
9. Schlichting, C.D., Pigliucci, M.: Phenotypic Evolution: A Reaction Norm Perspective. Sinauer Associates Incorporated (1998)
10. Kahneman, D.: Maps of bounded rationality: psychology for behavioral economics. Am. Econ. Rev. **93**(5), 1449–1475 (2003)
11. Conway, J.: The game of life. Sci. Am. **223**(4), 4 (1970)
12. Max Tegmark's answer to the Annual Edge Question (2017). https://www.edge.org/annual-questions
13. Fodor, J.A.: Special sciences (or: the disunity of science as a working hypothesis). Synthese **28**(2), 97–115 (1974)
14. Hoel, E.P., Albantakis, L., Tononi, G.: Quantifying causal emergence shows that macro can beat micro. Proc. Natl. Acad. Sci. **110**(49), 19790–19795 (2013)
15. Hoel, E.P.: When the map is better than the territory (2016). arXiv:1612.09592
16. Laplace, P.S.: Pierre-Simon Laplace Philosophical Essay on Probabilities: Translated from the Fifth French edition of 1825 with Notes by the Translator, vol. 13. Springer Science & Business Media (2012)
17. Pearl, J.: Causality. Cambridge University Press (2009)
18. Bateson, G.: Steps to an ecology of mind: collected essays in anthropology, psychiatry, evolution, and epistemology. University of Chicago Press (1972)
19. Tononi, G., Sporns, O.: Measuring information integration. BMC Neurosci. **4**(1), 31 (2003)
20. Shannon, Claude E.: A mathematical theory of communication. Bell Syst. Tech. J. **27**(4), 623–666 (1948)
21. Fodor, J.A.: A theory of content and other essays. The MIT Press (1990)
22. Maturana, H.R., Varela, F.J.: Autopoiesis and Cognition—The Realization of the Living, Ser. Boston Studies on the Philosophy of Science. Dordrecht, Holland (1980)
23. England, J.L.: Statistical physics of self-replication. J Chem. Phys. **139**(12), 09B623_1 (2013)
24. James, W.: The principles of psychology. Holt and Company, New York (1890)
25. Ashby, W.R.: An introduction to cybernetics (1956)
26. Marshall, W., Albantakis, L., Tononi, G.: Black-boxing and cause-effect power (2016). arXiv: 1608.03461
27. Kullback, S.: Information Theory and Statistics. Courier Corporation (1997)
28. Frisch, M.: Causal Reasoning in Physics. Cambridge University Press (2014)
29. Sperry, R.W.: A modified concept of consciousness. Psychol. Rev. **76**(6), 532–536 (1969)
30. Ellis, G.: How can Physics Underlie the Mind. Springer, Berlin (2016)
31. Noble, D.: A theory of biological relativity: no privileged level of causation. Interface Focus **2**(1), 55–64 (2012)

Chapter 7
Bio from Bit

The known laws of physics are impressive in their explanatory power. They describe for us the mundane, such as balls rolling down inclined planes, and the extreme such as what happens under gravitational collapse, or how the nucleus is bound together. However, so far even our best theories of physics are not yet capable of explaining why physical systems exist that can and do create theories to describe the world [17]. Arguably this is the most interesting feature of having laws at all.

A theory is a compressed description of the world, which is explanatory of some of its regularities. In short, theories are information. But this kind of information is not exactly what we commonly associate with the more widely discussed concept of information as presented originally by Claude Shannon. Shannon was interested in communicating information over a noisy channel, and therefore identified 'information' with reduction in uncertainty (Shannon 1949). The majority of work in information theory so far has similarly been focused on the idea of reducing uncertainty, or stated differently, maximizing the predictability of a given outcome. However, theories are powerful not only because they describe things, and thereby allow us to predict properties of the world, but because they also allow us to perform transformations in the physical world. Once a physical system, such as a living entity, overcomes the issues of noise and acquires information, that information is only retained if it is useful, that is if the system can do something because of it. Importantly, different transformations can occur depending on the information acquired. The capacity for information to cause transformations is most commonly discussed in the context of **'downward' causation**, where it is proposed information is an integral part of the causal structure of some physical systems [4, 8, 9, 22] (here causal structure implies the collection of what transformations can happen). This capacity to use information to perform specific transformations is a property not just of humans and our technology, but is also at the heart of what life does. Understanding how information

S. I. Walker (✉)
Beyond Center for Fundamental Concepts in Science, Arizona State University, Tempe, AZ, USA
e-mail: sara.i.walker@asu.edu

© Springer International Publishing AG, part of Springer Nature 2018
A. Aguirre et al. (eds.), *Wandering Towards a Goal*, The Frontiers Collection,
https://doi.org/10.1007/978-3-319-75726-1_7

can cause certain transformations to occur is critical to understanding life, and its goal-directed behavior.

7.1 Knowledge as Information with Causal Power

> Base metals can be transmuted into gold by stars, and by intelligent beings who understand the processes that power stars, and by nothing else in the universe.
>
> —David Deutsch [5]

If this were not the case, technological revolutions would not be coincident with scientific ones. An example is the launching of artificial satellites into space as an illustration of the process of "anti-accretion" [11, 23]. We are accustomed to the idea of the physical process of accretion, whereby planetary bodies acquire mass. Less often do we consider the inverse process of planets loosing mass, which can happen by 'natural' means, for example in the case of disintegrating planets, or by 'artificial' means, for example when objects, such as satellites, are launched purposefully to space.[1] Anti-accretion by the latter process requires as a pre-condition the existence of a technological civilization (e.g., on Earth this is humanity), and therefore happens on Earth but not on any other planet in our solar system. A further condition is the technological civilization must include *knowledge* of the regularities associated with gravitation. That is, stored somewhere in the physical degrees of freedom instantiating the civilization must be a theory of gravity. Knowledge in this sense is a special kind of information that includes the capacity to *cause* certain transformations to occur, in this case the launching of satellites to space (see e.g., Deutsch's and Marletto's work on Constructor theory, where a related concept of knowledge is proposed with the property it remains instantiated in physical systems after causing a transformation [7]). A physical understanding of knowledge is a necessary precondition for a theory accommodating the appearance of goal-directedness: in order to achieve a goal, one must have the knowledge to get there.

7.2 Macrostates Matter to Matter

It is often argued the idea of information with causal power is in conflict with our current understanding of physical reality, which is described in terms of fixed laws of motion and initial conditions. If deterministic physical laws describe all of reality, there is no "room at the bottom" for macro systems (such as living entities) to do causal work. The question of how mindless mathematical laws can give rise

---

[1] 'Natural' and 'artificial' are included in quotations as one primary point of this essay is that the distinction between natural and artificial (technological) systems may be an artificial one, after all technology physically exists too so it is reasonable to assume it is a natural outcome of the laws of physics (by no means proven).

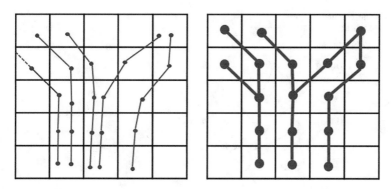

Fig. 7.1 Trajectories at the physical level (left) and at the emergent level of agents (right). Reproduced from [18]

to goal-directed behavior may therefore be recast in the framework of identifying how it is that macroscopic systems can appear to cause transformations, under the imposing constraint of deterministic laws. There are two possibilities, either (1) theories, theorizers and related processes exist at an emergent level of description and have no bearing on dynamics at the lower levels to which they supervene or (2) theories and theorizers exist and can potentially influence dynamics at lower levels.

I first address the former, which is how we typically regard emergence in physics and much of biology. A nice example is illustrated by List [18], reproduced in Fig. 7.1. The macrostate at any given instant in time M(t) is consistent with a set of microstates P, such that the macrostate is multiply realizable (this is the definition of a macrostate). Under deterministic evolution, a microstate $p \in P$ could lead deterministically to a new microstate $p' \in P'$, where P' is the set of microstates consistent with a new macrostate M'(t + 1). Macrostates of interest in biology are those that have agency, that is that have 'causal power'. Agents can exist among many counterfactual histories, imbuing them with their seemingly ability to cause different counterfactual possibilities to occur (see e.g., [3] for discussion).

The left panel of Fig. 7.1 shows examples of macro trajectories that bifurcate in the future, and trajectories that bifurcate in the past. The possibility of multiple possible pasts, each equally consistent the same current state, leads to the appearance of goal-directed behavior in macroscopic systems if we impose assumptions that there is only one possible real past history (i.e., it is deterministic). Cosmologically, we often will conceptually rewind the tape of the history of the entire universe all the way back to its initial state to gain this kind of explanatory power. This leads to problems associated with fine-tuning the initial state of the universe to explain the world as we see it. However, if we only care about macroscopic systems, they can have multiple possible histories. Assuming only one history in a case where there are many leads to the appearance of design and fine-tuning. Looking at all possible counterfactual paths to achieve a given target state we should observe that "the goal" (i.e., the current state of the system, or a specified target state) is a possible outcome arising due to

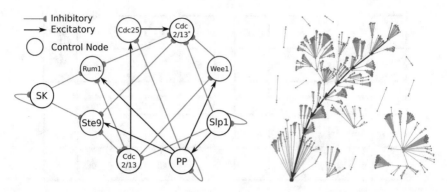

Fig. 7.2 The fission yeast cell cycle regulatory network modeled as a Boolean network (left) and the state-transition diagram generated by the network (right). Figure adopted from [19] (left) and [6] (right)

the convergence of many different trajectories (e.g., the rightmost trajectories of the right panel) and does not require a unique initial condition (e.g., does not need to be "designed").

Goal-directedness is closely connected to the idea of biological function. In systems biology, function (and by extension goals) are often modeled within the framework of attractor landscapes (see e.g., [14]), where attractors may be regarded as "goal-states" and the "function" of a biological network is to drive a given system's dynamics to specified attractors that correspond to viable phenotypes [15]. An explicit example is the attractor landscape for the Boolean network model of the fission yeast cell cycle gene regulatory network [6], shown in Fig. 7.2, which models the function of cellular division by recapitulating the sequence of steps executed for a subset of genes in living *Schizosaccharomyces Pombe* cells. The network in Fig. 7.2 is a coarse-grained representation of what we think actually happens inside the cell. It can be thought of as the product of a mapping, or coarse-graining of cellular function, much like the coarse-graining of Fig. 7.1 (where the cell cycle model would represent trajectories in the right panel). Here the Boolean states of nodes correspond to whether a given gene (node) in the network is expressed ('1') or not ('0') (often in Boolean networks individual nodes represent multiple genes and therefore are a true coarse-graining of the dynamics). The trajectories for all possible initial states of the network are shown in Fig. 7.2. The sequence of blue arrows corresponds to the healthy cell cycle trajectory, such that each state along the path corresponds to one of the phases of cellular division (G1, S, G2, M corresponding to growth, replication, a resting gap state and mitosis, respectively). The end state is the primary attractor for a healthy cell, and it is clear from the state transition diagram that this attractor dominates the landscape. Most stages converge to the primary attractor, and for this reason the cell cycle is said to have robust function. This is a product of evolutionary selection, which has encoded in this particular network the function of converging on this particular state as a viable resting state for the cell (attractor).

There is a lot more information in the biosphere today than there was at the time of the last universal common ancestor ~3.5 billion years ago. Presumably, this trend is driven by evolution encoding more information in biological networks over time through selection. Observed over long timescales the macrostates that persist are predictive [10], and were selected for the precise reason they have staying power (are correlated with future states). But, there seems to be no steadfast rules about which of the vast set of possible macrostates are physically relevant. Assuming a state of n elements, the number of possible partitions of the state into sets of elements is given by the nth Bell number $B_n = \sum_{k=0}^{n} \binom{n}{k}$. This is huge, even for small n. Yet, only some macrostates appear to persist in the world (in general, you are at least somewhat the same person you were yesterday, indicating something about your macrostate persisted). This can be explained if the relevant macrostates are good predictors of future dynamics, which is really just a statement that the past is correlated with the future. In statistical physics for example, we care about the macrostates that are optimal predictors of their own future evolution in time, and there is a unique minimal, objective solution for this (Shalizi and Crutchfield 2001). But, observing macrostates is also subjective and requires introducing an external 'observer' (Shalizi and Moore 2003) (implying there is an external system doing the coarse-graining). Ideally, we should be able to explain the properties of biology without assuming an external system is generating predictions.

The foregoing discussion of coarse-graining and multiple-histories is unsatisfactory. It falls short of answering our original question: why is it some macroscopic configurations of matter (i.e., us) can construct theories about their world? Not all information encoded in the biosphere is necessarily predictive. We can go back to the satellite example. Certainly, part of the power of knowledge of the laws of gravitation is that we as a species can now predict the path of the Moon and other planets. But more significant is that we can use this knowledge to launch satellites, build GPS devices and develop other technological innovations that lead to further innovation. That is, this knowledge allows for multiple counterfactual possibilities to be realized, which originate at the 'level' the knowledge is instantiated (branching in right of Fig. 7.1). This introduces asymmetries in how we treat different levels in a hierarchy (here so far, I have been talking about 'macro' and 'micro', but this should always be assumed to include a hierarchy of nested levels across scales). As currently understood in physics, counterfactuals can matter for the higher levels (e.g., via the branching in Fig. 7.1) but not the lower levels, as lower levels follow local time-reversal invariant laws. We therefore can solve problems regarding fine-tuning and permitting goal-directedness at some levels (the macro ones) but not all (the micro ones). It may be that nature is structured this way, but there is also no a priori reason to assume that we should not treat all levels equally (for example, there may be no bottom level rendering it impossible to treat a particular level as privileged [21]). Adopting the view that macrostates can matter, resolves many of this issues and provides a path for explaining why theories exist in the first place—they enable more transformations to occur than would be possible otherwise.

7.3   More Is Not Just Different, It Is Causal

Our second possibility is theories and theorizers can potentially influence dynamics at lower levels. Macrostates may not just be predictors, viz the informational properties of emergent agents, but could also play a role in the *causal structure* of reality. When studying complex systems it is evident that not all of reality may be reduced to description solely in terms of those fundamental laws that operate at its most microscopic layers. In the words of the Nobel Laureate Philip Anderson in his famous essay *More is Different* [2]:

> The ability to reduce everything to simple fundamental laws does not imply the ability to start from those laws and reconstruct the universe.

As we move up in scale from subatomic particles, to atoms, to molecules, to life and mind, the universe appears to become more and more complex, and it seems clear as complexity increases entirely new properties emerge. The question of interest is whether these emergent properties can themselves have causal consequences, allowing agents to have an active say in achieving their own goals. A point often discussed but still under-appreciated is emergence and reductionism are not at odds. In fact, reductionism is what makes emergence possible: if we could not reduce reality to the study of a few component building blocks, we would not be able to describe how those building blocks come together to create more complex 'higher-levels'.

Hoel et al. explicitly defined a concept of *causal emergence* by identifying *cause* and *effect* information, quantified by determining how the mechanisms of a system constrain its past and future states when the system is in a particular state [12]. They show that even though there may exist a complete (causally closed) lower-level description of a physical system, higher levels can still be causal above and beyond the causal work of lower levels if higher levels have higher cause-effect information, that is if they are more deterministic or more specific in their dynamics (see [12, 13] for description).

Cause-effect information is important in integrated information theory (IIT) as developed by Tononi and collaborators, where integrated information ϕ provides an explicit measure of how a whole can be "more than the sum of its parts" [20]. Briefly, a system is integrated if when cut in parts it looses cause-effect power. Recently, my collaborators and I performed causal analysis on the fission yeast cell cycle regulatory network Boolean model in Fig. 7.2 using IIT [19]. The analysis uncovered the cell cycle in its primary attractor state is an integrated whole, as reproduced in Fig. 7.3, where local maxima of cause-effect power are shown outlined in blue-dashes. The most irreducible structure (the global maximum) is the cause-effect structure including all nodes of the cell cycle, exclusive of SK (the start node, which has no other feedback on the network). The analysis also revealed a local maximum of cause-effect structure, which corresponds to the set of nodes that when individually intervened on (by setting the state to their value in the primary attractor state at each step) expand the size of the primary basin of attraction and therefore increase its functional robustness (where in the systems biology of attractor landscapes, robustness is typically associated with the size of the basin of attraction for a given phenotype). In other

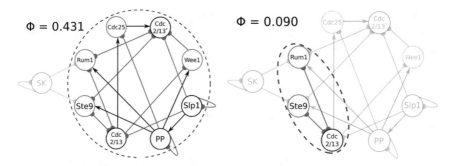

Fig. 7.3 Local maxima of intrinsic cause-effect power in the primary attractor of the fission yeast cell cycle network (left). The global maximum corresponds to the full network (with the exception of the SK which initiates the cycle). A second local maximum corresponds to nodes that individually enhance the function of the network by expanding the size of the primary basin of attraction (right). Figure reproduced from [19]

words, these nodes increase the goal-directed behavior of the network. The network was also shown to be integrated as a whole (all 8 nodes excepting the 'start' node SK) through each state of the biological attractor. This network therefore provides an example of a system that maintains causal borders as an integrated entity in executing a goal-directed function (converging to an attractor).

The causal analysis of the Boolean fission yeast cell cycle network was also compared to the 'backbone motif' of the network [24]. The backbone motif is defined as the minimal subgraph of the network in the left panel of Fig. 7.2 that still recapitulates the sequence of states along the blue path in the right panel of the figure. In short it is a smaller network that has the same function as the full cell cycle network. Causal analysis reveals this network does not form an integrated whole, and looses integration at various stages of the cell cycle process. This is important for understanding how biological systems can wander towards a goal—it is not a goal if it is not internally "programmed". For the backbone motif it is almost coincidence (but for the fact that it was designed from a biological circuit) the desired target state is achieved. The basin of attraction for the primary attractor of the backbone motif network is much smaller than it is for the original fission yeast cell cycle network. For the full cell cycle network, integration is critically important as it sets the causal boundaries of "self". There are subsets of nodes within the network (those corresponding to the local maxima in the right panel of Fig. 7.3) regulating its function to achieve the targeted state. This underlies its robustness.

The forgoing example of the cell cycle network highlights an important feature of the causal structure of biological systems—they are integrated containing many "higher-order" (meaning sets of nodes here, rather than just individual nodes) cause-effect structures that *control* their own function. **One way to think about living structures is as a nexus of counterfactual possibilities, such that a living organism not only exists across multiple trajectories, but is an embedded causal structure that itself regulates which paths are taken**. Agency and goal-directed behavior are an emergent property arising due to the dynamics of these highly embedded structures through state space, arising because they define their own trajectory through regulatory feedback. One final example will help illustrate this point.

7.4   Life as 'Non-trivial' Trajectories

To illustrate how 'life' can alter the causal structure of dynamical systems it is embedded via coarse-graining, I will use as an example Elementary Cellular Automata (ECA). ECA are 1-dimensional with nearest-neighbor interactions operating on a two-bit alphabet, here represented as $\{0, 1\}$. The topology of the state transition diagram for ECA Rule 150 with width $w = 6$ and open boundaries ('0' on the boundary) is shown in the left panel of Fig. 7.4. Here I focus on Rule 150 as it is reversible, such that the state-transition diagram on the left of Fig. 7.4 looks like what we would expect in a physical system (that is if it were local, deterministic and time-reversible, which is a standard assumption for most of physics). To introduce coarse-graining as an endogenous property of the CA, it is partitioned into two parts, each with size $w = 3$. Each partition receives as an input only the coarse-grained state of the other. All possible transitions arising from all possible partitionings of the state space into two 'coarse-grained' bins are calculated, subject to the constraints that '000' and '111' do not appear in the same partition and each partition includes at least two states. There are 62 possible partitions satisfying this constraint, meaning there are 62 possible ways for each CA to 'coarse-grain' the other. Shown on the right, is the resulting state transition diagram with this kind of explicit 'downward' causation introduced. Different edges in the graph correspond to different transitions the system can take depending on how the two CA are coarse-graining one another. Such a mechanism could operate in the real-world if physical systems really can interact with the external world via a reduced number of degrees of freedom (a not unreasonable assumption). One example of where coarse graining is apparent in biochemistry is the mapping of the genetic code, which includes a reduction in information of 3 bits to 1bit in mapping from nucleic acids to protein. Physical systems can have a fixed partitioning of the state-space of their environment that sets their interactions for all time, or coarse-grainings can evolve in time as has occurred through biological evolution.

More is different: the dynamics you get for the same local rule under partitioning is much richer than to what you get without it. The path-lengths of all trajectories for the state transition diagrams in Fig. 7.4 are shown in Fig. 7.5. In blue are the trajectory lengths for all $w = 6$ ECA (not just rule 150). In red are shown the trajectory lengths for all partitioned CAs interacting via coarse-graining, where the only rule used is rule 150. By adopting an explicit framework whereby systems have borders and are coupled to the external world via macrostates (information channels) richer dynamics are possible. In fact, simpler rules sharing properties consistent with the physical world (local, deterministic and reversible) can become explanatory of much more

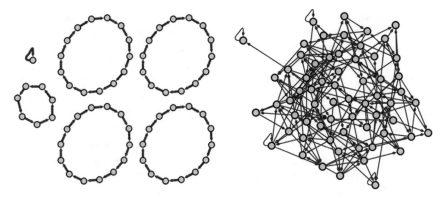

Fig. 7.4 State transition diagram for ECA rule 150 (left) and for rule 150 under all state bipartions as described in the text (right) for CA of width w = 6 and open boundary conditions

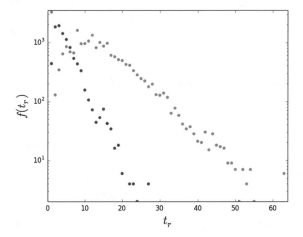

Fig. 7.5 The frequency distribution of all possible trajectory lengths taken over all state-transition diagrams for the 256 ECA rules (blue), compared to the frequency distribution of trajectories for Rule 150 for all partitioned CAs interacting via coarse-graining

complex emergent dynamics if interactions via macroscopic states, as introduced here, are possible.

The main goal of this essay is to suggest life has an underlying state-transition diagram much more like the right of Fig. 7.4 than the left (and is not merely a coarse-grained macrostate of the left, see also Adams et al. [1]). Living systems are information processing hierarchies [10], that through nested levels coupled by "information" can achieve trajectories impossible to describe with a fixed law and initial state only (where the latter are a subset of the former). A biological system constructs its path through these high dimensional spaces, highly integrated structures are more adept at this, and goal-directed behavior emerges because each state containing 'life' (coupling interactions across scales) has multiple possible futures and pasts.

Explicitly introducing something as odd as macro-level causal effects may seem to some to be premature. We may perhaps find a more conventional description of living matter without having to uncover new principles to explain how macro can interact with micro (again, here I talk of two levels but there is always a hierarchy of many scales corresponding to all the ways a system's state space can be partitioned). My goal is not to make claims one way or the other, but instead to explicitly study the consequence of taking the idea of downward causation as a physical effect seriously and study its dynamical consequences. However, it should be noted that the models presented need not be interpreted as a mysterious or 'action-at-a-distance' to explain downward causal effects or embedded causal structures. There can exist consistent micro-narratives for the dynamics that do not include explicit downward causation per se, but instead require adopting a state-dependent topology for the CA (which can be defined entirely locally), rather than the fixed lattice topology of traditional ECA. This requires the rules of interaction change with time in a manner dependent on the state of the system. Since the laws of physics appear to be immutable the freedom to do this must come from the interaction topology. State-dependent dynamics are frequently cited as a potential hallmark feature of life (being self-referential) ([1]; Goldenfeld and Woese 2011), and it has been proposed elsewhere that life might implement state-dependent topology in its dynamics (Rosen 1981). A consequence is that when we describe networks in biology, we are really gaining a window into the complex network that is itself the causal structure reality. There is no underlying local, fixed dynamical law narrative that will describe fully the longterm dynamics of biological or technological systems even if every step is consistent with the laws individually (hence even though Rule 150 is used for all transitions in Fig. 7.4, the two diagrams look different). Here we see a new interpretation of Wheeler's famous dictum "it from bit". The participatory universe exists due to modifications to the causal structure of reality in the vicinity of observers: it is how information restructures the physical world that enables the "bio to emerge from the bit".

References

1. Adams, A.M., Berner, A., Davies, P.C.W., Walker, S.I.: Physical universality, state-dependent dynamical laws and open-ended novelty. Entropy **19**(9) (2017). https://doi.org/10.3390/e19090461
2. Andersen, P.: More is different. Science **177**(4047), 393–396 (1972). https://doi.org/10.1126/science.177.4047.393
3. Biehl, M., Ikegami, T., Polani, D.: Towards information based spatiotemporal patterns as a foundation for agent representation in dynamical systems. Proc. Artif. Life Conf. **2016**, 722–729 (2016). https://doi.org/10.7551/978-0-262-33936-0-ch115
4. Campbell, D.T.: 'Downward Causation' in Hierarchically Organised Biological Systems. Studies in the Philosophy of Biology, pp. 179–86 (1974). https://doi.org/10.1007/978-1-349-01892-5_11
5. Deutsch, D.: The Beginning of Infinity: Explanations that Transform the World (2011)

6. Davidich, M.I., Bornholdt, S.: Boolean network model predicts cell cycle sequence of fission yeast. PLoS ONE **3**(2) (2008). https://doi.org/10.1371/journal.pone.0001672
7. Deutsch, D.: Constructor theory. Synthese **190**(18), 4331–4359 (2013). https://doi.org/10.1007/s11229-013-0279-z
8. Ellis, G.F.R.: Top-down causation and emergence: some comments on mechanisms. Interface Focus **2**(1), 126–140 (2012). https://doi.org/10.1098/rsfs.2011.0062
9. Flack, J.C.: Coarse-graining as a downward causation mechanism. Philos. Trans. R. Soc. A Math. Phys. Eng. Sci. **375**(2109), 20160338 (2017). https://doi.org/10.1098/rsta.2016.0338
10. Flack, J.C., Erwin, D., Elliot, T., David, D.C., Krakauer, C.: Timescales, symmetry, and uncertainty reduction in the origins of hierarchy in biological systems. Coop. Evol. 45–74 (2012). http://scholar.google.com/scholar?hl=en&btnG=Search&q=intitle:Timescales+,+Symmetry+,+and+Uncertainty+Reduction+in+the+Origins+of+Hierarchy+in+Biological+Systems#0
11. Grinspoon, D.: Earth in Human Hands: Shaping Our Planet's Future (2016)
12. Hoel, E.P., Albantakis, L., Tononi, G.: Quantifying causal emergence shows that macro can beat micro. Proc. Natl. Acad. Sci. **110**(49), 19790–19795 (2013). https://doi.org/10.1073/pnas.1314922110
13. Hoel, E.P.: When the map is better than the territory. Entropy **19**(5) (2017). https://doi.org/10.3390/e19050188
14. Huang, S., Eichler, G., Bar-Yam, Y., Ingber, D.E.: Cell fates as high-dimensional attractor states of a complex gene regulatory network. Phys. Rev. Lett. **94**(12), 1–4 (2005). https://doi.org/10.1103/PhysRevLett.94.128701
15. Kauffman, S.: The Origins of Order: Self-organization and Selection in Evolution (1993)
16. Kim, H., Davies, P., Walker, S.I.: New scaling relation for information transfer in biological networks. J. R. Soc. Interface **12**(113) (2015). https://doi.org/10.1098/rsif.2015.0944
17. Krakauer, D.C.: The Complexity of Life. Santa Fe Institute Bulletin (2014)
18. List, C.: Free will, determinism, and the possibility of doing otherwise. Nous **48**(1), 156–178 (2014). https://doi.org/10.1111/nous.12019
19. Marshall, W., Kim, H., Walker, S.I., Tononi, G., Albantakis, L.: How Causal Analysis Can Reveal Autonomy in Models of Biological Systems. Philos. Trans. R. Soc. A Math. Phys. Eng. Sci. **375**(2109), 20160358 (2017). https://doi.org/10.1098/rsta.2016.0358
20. Oizumi, M., Albantakis, L., Tononi, G.: From the phenomenology to the mechanisms of consciousness: integrated information theory 3.0. PLoS Comput. Biol. **10**(5) (2014). https://doi.org/10.1371/journal.pcbi.1003588
21. Schaffer, J.: Is there a fundamental level? Noûs **37**(3), 498–517 (2003). https://doi.org/10.1111/1468-0068.00448
22. Walker, S.I.: Top-down causation and the rise of information in the emergence of life. Information (Switzerland) **5**(3) (2014). https://doi.org/10.3390/info5030424
23. Walker, S.I.: The descent of math, 183–92 (2016). https://doi.org/10.1007/978-3-319-27495-9_16
24. Wang, G., Chenghang, D., Chen, H., Simha, R., Rong, Y., Xiao, Y., Zeng, C.: Process-based network decomposition reveals backbone motif structure. Proc. Natl. Acad. Sci. USA **107**(23), 10478–10483 (2010). https://doi.org/10.1073/pnas.0914180107

Chapter 8
I Think, Therefore I Think You Think I Am

Sophia Magnúsdóttir

Conceptual clarity is the foundation of scientific discourse. Therefore, I wish to propose a new way to speak about and quantify consciousness. This new definition is based on the ability of a system to accurately monitor and predict its environment and itself. While I am at it, I will also explain philosophical zombies, free will, and the purpose of life.

8.1 The Big, Bad Question

"What is consciousness?" isn't only a big question, it's also a bad question. That's because any consistent definition would answer it. Here is one: "Consciousness is a lemon tree." That, you might complain, is not what you mean by consciousness. Too bad, then, that you cannot tell me what you mean because you don't know what consciousness is. What a mess.

8.2 The Goal

To make headway on this big, bad question, I therefore first have to make the question more concrete. I am looking for a definition that captures what most of us (humans, presumably) mean by consciousness, a definition according to which (a) rocks aren't conscious but (b) most animals are, (c) animals can be conscious to varying degrees, and (d) even be temporarily unconscious. Besides fulfilling the requirements

S. Magnúsdóttir (✉)
University of Gothenburg, Gothenburg, Sweden
e-mail: sabine.hossenfelder@gmail.com

© Springer International Publishing AG, part of Springer Nature 2018
A. Aguirre et al. (eds.), *Wandering Towards a Goal*, The Frontiers Collection,
https://doi.org/10.1007/978-3-319-75726-1_8

(a)–(d), the definition should also enable us to answer following three representative questions:

(Q1) Is an anesthetized person safely out so that they do not experience pain?
(Q2) Is a person with locked-in syndrome self-aware and/or aware of their situation?
(Q3) Has an artificial intelligence developed consciousness comparable to that of animals?

Finding a definition that fulfils requirements (a)–(d) and allows answering questions Q1–Q3 is surprisingly difficult. The most common way to test for consciousness is to seek a reaction, for example by prodding a patient or checking pupil contractions. But this probes for more than just consciousness because it moreover requires visible output.

We might want to assert that certain reactions are, if not necessary, then at least sufficient indicators for consciousness. This assertion is questioned by the concept of a "philosophical zombie," an imaginary being that reacts like a human yet does not have experience.[1] But is it even possible for someone (some thing?) to behave like a human without also having human-like experiences? This is another question which a good definition of consciousness should be able to settle.

A somewhat more advanced test for consciousness might scan brain activity. But this does not help us much either because we don't know which type of brain activity demonstrates experience or consciousness. Again, we might want to search for a reaction to stimulus, but any computer with an input/output (hereafter I/O) interface, such as a camera or simply a keyboard, does—in some sense–'react' to input. Most of us, however, do not ascribe consciousness to present-day computers.

Previously proposed output-independent measures for consciousness have focused on a system's capability for and type of information processing, like Tononi's Integrated Information Theory[2] or Tegmark's Consciousness as a State of Matter.[3] These approaches suffer from two problems. First, it isn't clear whether they actually capture what we mean by consciousness, but—in all fairness—these are early days for consciousness models and settling the issue might just take more time. More importantly, however, approaches that aim to quantify consciousness based on a system's structure leave us wondering what the system is supposedly conscious of.

So, yes, it's a hard problem, but if it was easy then what would be the point of this contest? The goal of this essay is hence to answer the above questions. Since I am not a neurologist, however, the reader be warned that I only offer an answer 'in principle,' not an answer 'in practice.' I believe, however, that with further refinement the approach I want to propose can become practically useful (see Appendix).

8.3   Clues from Evolution

Let us begin by asking how we even got to the point of asking "What is consciousness?" We are products of Darwinian evolution. 'Survival of the fittest' is commonly interpreted as an adaptive selection of actions beneficial for reproduction. But this pays too little attention to the question what it takes to develop these reactions.

Take, for example, a rabbit. A rabbit which runs or hides when it sees a tiger has an evolutionary advantage over a rabbit that mistakes the tiger for a carrot—that much is clear. Less clear is what the rabbit brain must do to recognize the threat. (I'm not sure there are many rabbits in the natural habitat of tigers, but you get the point.)

To begin with, the rabbit brain must be able to obtain sufficiently accurate information about its environment, which means it needs an input channel. More importantly for our purposes, it must be able to identify patterns matching the threat 'tiger.' This means the rabbit brain must be able to create a reasonably accurate model of its environment: It must have an internal representation that faithfully encodes information about the environment, and further a way to process this representation to arrive at a reaction.

This internal representation allows for reliable if-then reactions, which is a good starting point. But the rabbit can become even 'fitter' when it uses more sophisticated models, which take into account not only the present state of the tiger but extrapolate the tiger's behavior, possibly including also the reaction of the environment. Does the tiger look like it has seen the rabbit and is about to jump? If it jumps, where will it likely land? What's the tiger-ground friction coefficient? If the rabbit has a predictive model that computes faster than the environmental story unfolds, it will allow the rabbit to get out of the way. Accurate predictions, therefore, are plausibly a survival advantage and hence a likely product of evolution.

Indeed, there is much evidence that the brain is good at exactly these tasks. Our ability to extrapolate trajectories seems at least partly genetic. Mouse brains are known to have 'place cells' corresponding to their location in a maze[4] and various species have been found to have brain structures encoding small integers.[5] All these are examples of environmental models. Consequently, pattern recognition is what neural networks are now trained to perform at, in mimicry of human development.

8.3.1   Terminology

Let us be precise with the terminology by employing the language of mathematics. One may debate whether consciousness can be entirely captured by mathematics, but this need not worry us here. We merely note that mathematics has been successfully applied to categorize and understand many natural systems, and it therefore seems reasonable to also use it to better classify consciousness.

By **system** we will simply mean a set with distinct elements that have properties and relations among each other. A **subsystem** is a subset of the system. We'll refer to the largest possible system as **universe**. If the system under consideration is not identical to the universe, we call its complement **environment**. Remaining connections between a system and the environment represent input/output channels.

Note that in these definitions we have not made any assumptions about the spatial relations between the elements or their compactness. We will come back to this later.

The reductionists among the readers may want to think of the system's elements as elementary particles and the relations as interactions, but there is no need for this particular interpretation. The system's elements can as well be emergent objects in a higher-level theory. However, I wish to emphasize that it is not necessary to identify the elements of the system with synapses connected by axons, they could as well be circuits or qubits.

We will use the word **model** to mean a morphism, that is a map between elements which preserves the relation between the elements and the assignment of properties. Ie, two elements that have the same property (relation) P in the origin have the same property (relation) P' in the image, though two elements that had different properties (relations) P and Q might end up having the same property (relation) P'. In other words, the morphism is not, in general, an isomorphism.

Since nothing real is ever perfect, the morphisms we will deal with here generally have a limited fidelity. By this we mean that the image of the morphism doesn't preserve the relations or properties of all elements correctly. We will hence speak of **imperfect morphisms**.

The fidelity of the morphism quantifies the model's accuracy. There are various ways to quantify fidelity—we could for example just count the number of mistakes in the properties and elements, and divide them by the number of properties and elements. There is no unique definition for fidelity but for our purpose we only need to know that it is a property which *can* be quantified. That such a quantification isn't unique means that the absolute value of a model's accuracy isn't meaningful, but we can still make relative comparisons. (Similar to the case with the utility function employed in equilibrium economics.)

The system and the environment will further in general be dynamical, meaning they will change over time. If the morphism is also time-dependent but achieves to model a time-sequence of states of the environment faster than the change in the environment occurs, we will refer to it as a **predictive morphism**. Again we could quantify how good the model's prediction is by measuring the mistakes that occur over some time.[6]

8.4 Four Levels of Awareness

Armed with this terminology, let us return to the rabbit. We have seen that a system has better chances of survival when it has an accurate model of the environment, better still if the model is moreover predictive. The rabbit only benefits from this if it is able to do anything in reaction, like run away, but the model is a prerequisite for this.

The prediction of unfolding events, however, will be even better if the rabbit takes into account also its own actions and their likely consequences. Therefore, the rabbit increase its evolutionary advantage if it understands its likely reaction to a certain input. In other words, the rabbit will perform better if its brain has a model of itself modeling.

This brain self-model can, again, either merely monitor the rabbit's modeling—a non-predictive morphism—or it may be predictive. The former case we will refer to as 'experience,' the latter as 'cognition.' Finally arriving at the concept of consciousness, we will refer to the two cases which include a self-model as 'conscious awareness,' whereas the predictive and non-predictive models without self-monitoring represent 'unconscious awareness.'

This leads us to the following four-level classification of consciousness (see Figure left):

The 4 Levels of Awareness

Unconscious	**Conscious**
Perception Unmonitored, non-predictive morphism	*Experience* Monitored, non-predictive morphism
Projection Unmonitored, predictive morphism	*Cognition* Monitored, predictive morphism

1. **Perception**

The system receives input and produces a model of its environment. This model will in general have a limited fidelity. The better the fidelity the higher the awareness of the environment.

2. **Projection**

The system has a predictive model of its environment. Predictive models are only useful if the prediction of an event can be made before the event happens.

3. **Experience**

The system contains a subsystem with a self-similar model of itself that monitors the integration of input and its internal connectivity possibly including the monitor itself. This is akin the task-manager of an operating system.

4. **Cognition**

A system has cognition if it contains a subsystem with a self-similar predictive model of itself, again possibly including the monitor itself. (See Figure below.) The self-model, importantly, not merely again images the environmental model, but must also model the connections between the subsystems that models (a certain part of) the environment and other subsystems (dotted).

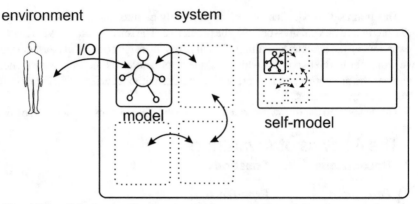

The ability of the human brain to model its own self-modeling is presently poor. This is the very reason we're having this essay contest.

8.5 Learning from that

An immediate consequence of the above proposed definition is that a system isn't either conscious or not, but it can be conscious to varying degrees. Consciousness, hence isn't binary but continuous. How conscious a subsystem is of its environment or other subsystems, depends on how good a model it has, how predictive the model is and how well it is monitored. The fidelity of the morphisms therefore quantify conscious and unconscious awareness.

A more important lesson, however, is that consciousness is not in and by itself a property of a system. Instead, consciousness is relational: Its origin lies in the relation between a system, its environment, and its subsystems.

Consciousness, hence, is a noun that is shorthand for a verb much like, for example, the word "leadership". Leadership too isn't a thing and it isn't a property, it's a relation. You can't just lead, you can only lead somebody or something. Consciousness, likewise, isn't a thing, it's a relation. You're not just conscious, you are always conscious *of* somebody or something.

It is well possible that the structurally-based models of consciousness like the ones proposed by Tononi and Tegmark identify exactly the kind of systems which can achieve exactly the model-building discussed above here. I believe, however, that focusing on self-similar maps is a more minimalistic way to think of consciousness.

A further lesson from the above is that consciousness is extendable: Since consciousness is the ability of a system to comprehend another system, consciousness can be expanded by supporting this comprehension externally. Moreover, as anticipated above, there's no particular reason why a conscious system needs to by compact or consist of components close by each other to create a predictive model. The components should be connected so as to exchange information, but other than that consciousness could extend over long distances. However, we may suspect that information-processing in very extended systems will be too slow and inefficient so that the evolution of consciousness is unlikely.

A point which we have not addressed above is the irreducibility of the conscious system. According to the above, any system that contains a conscious subsystem is also conscious, which is clearly not the way that we think about consciousness. Therefore, we should complete our definition by requiring that removing any elements or connections from the system will significantly decrease the system's level of consciousness (fidelity of the model).

To be more concrete on this point we would first have to distinguish connections between elements which carry information from those which carry supply (say, blood). At present, however, it isn't possible to clarify this definition because nobody knows how much of cognition may be 'embodied.' Indeed the definition proposed here could aid in identifying embodied cognition.

In summary, we have learned that consciousness is continuous, relative, extendable, and distributable.

8.6   Self-check

Let us now see whether we met the goal we set for ourselves.

(a) Rocks rarely change internal states, hence cannot create models of their environment, at least not in the typical lifetime of solar systems. A simple self-similar system can, to some extent, be said to have an experience of itself, but so long as it's not an experience of anything in particular, we probably wouldn't ascribe it much meaning. Having said that, it's interesting to note that self-similarity and scale-invariance are indeed hallmarks of complex systems and a relation to consciousness has been suggested previously.[7]

(b) Animals can, to varying degree, model their environment as evidenced by their ability to react and can therefore safely be said to be aware of their environment. The typical test for self-awareness—that an animal is able to recognize its reflection in a mirror—also beautifully fits into our definition. Recognizing one's reflection means that one has been able to identify part of the environment with internal processes, which requires self-monitoring and integration of one's own environmental modeling with the monitor.

(c) Moreover, animals can lose consciousness to a certain degree if their brains work in different ways, either because of malfunction or sleep, in which case their ability to model the environment and/or themselves is reduced.

We have hence fulfilled requirements (a)–(c).

Present-day robots have low levels of consciousness, if any. They are able to perceive parts of their environment, and may be programmed to react to it, but their models are primitive and, so-far, mostly unpredictive. The same can be said about the present status of artificial intelligences. Deep learning, however, holds much promise because it allows a system to anticipate what is going to happen. Our definition implies that that to advance artificial intelligence, neural networks should preferably be fed with time-ordered sequences of events for only this will enable them to develop predictive models and, thereby, consciousness.

Based on the definition proposed above, we could, on principle attempt to scan all possible subsystems of a human brain, animals, or computers, and compute the fidelity of the subsystems with parts of the environment or other subsystems. This means, in principle we can answer Q1–Q3. However, even if it was possible to monitor a brain with sufficient accuracy this task would presently be computationally infeasible. In the Appendix, I therefore propose how a presently possible practical test of the definition proposed here could look like.

What about the philosophical zombies? Can a system without experiences behave as if it was human? Yes, it can. The zombie's head could be empty except for a random generator that outputs arbitrary actions. By pure coincidence, the zombie might then behave exactly like a human and yet not have anything resembling experience according to our definition. This isn't impossible—but it is extremely unlikely. Indeed, as we have argued above, experience is hugely beneficial for the evolution of complex behavior that signals the system's ability to predict itself and its environment. Experience, hence, isn't necessary but likely to be found in any system able to behave even remotely human-like and philosophical zombies extremely rare.

In conclusion, we see that the trademark of consciousness is not that we think, but that we think about ourselves, about what others think, and what they think about what we may be thinking. We also, increasingly think about our own thinking, which might well be the path to higher consciousness.

8.7 Goals, Purpose, and Free Will

Let us then move towards derived concepts.

As previously noted, the ability of a system to predict the time-evolution of itself and the environment is limited because the more accurate a model, the more computationally intensive it becomes. This also means that generically a system can't predict its own reactions with absolute accuracy. This is why we have the impression of free will: Regardless of whether or not the future is indeed pre-determined, our inability to be sure about what we ourselves will do implies our self-models will project different possible future evolutions.

The trade-off between computational speed and accuracy also means that a system's ability in developing models will evolve towards a sweet spot where added computational effort for more accurate predictions is disfavored because the system's reaction would come too late to still bring more benefits.

Particularly valuable are hence models that have a low computational effort and yet are highly predictive. This is what goals are. We say that a system has a certain state (of itself and/or the subsystem) as "goal" if the optimization of actions to obtain the goal is predictive of the dynamics of the system.

Note that it isn't necessary for the system to actually reach the goal for it to be predictive! The goal of winning this essay content, for example, predicts that I agonize over every single word, regardless of whether I win.

We do not normally speak of 'goals' when referring to non-conscious systems, so to better match the common use of the word we could restricted this notion to conscious systems. Otherwise the moon could be said to have the goal of falling onto Earth, just that its initial velocity prevents it from ever reaching this goal. I leave it up to you whether or not you might want to add this restriction to the definition.

The existence of humans demonstrates that, given enough time, systems can develop remarkable predictive abilities and complicated reactions which in return are difficult to predict. Survival and reproduction, therefore, which once were highly predictive, might cease to be predictive in the long run. And so, when we start with assuming a goal to predict what happens in the course of evolution, we have it exactly the wrong way round: It is instead the possibility to make predictions with it that defines what we mean by goal.

Having come so far, it is straight-forward to also define what we mean by 'purpose.' We say an object or action has a certain 'purpose' if it increases the likelihood of a system reaching a goal. The word is also sometimes used to mean what we might call more descriptively 'intended purpose.' This is to mean a system might have computationally arrived at the conclusion that an action serves the purpose of reaching a goal, but the conclusion might be an inaccurate.

Humans are guilty of applying terms out of context, thus we often speak in extension of the purpose of abstract concepts other than objects or actions, for example the purpose of life. This is another big, bad question, to begin with because life, too, is not presently a well-defined notion. But even assuming that we had a good definition for life, speaking of its purpose would require us to first identify a goal that life might contribute to reaching.

Hydrogenating carbon dioxide, then, is not such a bad guess[8] for a goal that we could attribute to the universe, though it doesn't seem hugely predictive for the finer details. Based on the above we could instead conjecture that a more predictive goal for the evolution of the universe is becoming an accurate and predictive model of itself. The purpose of life, then, would be to develop these models. In other words, the purpose of our existence might be understanding the universe, but I admit that at this point I have succumbed to insubstantial speculation.

But we can now also ask what is the purpose of a specific life, yours or mine or the rabbit's. Let's take my life as example because it's the one I can speak about most confidently. To find the purpose of my life I must first ask which goal I am contributing to. Again, hydrogenating carbon dioxide rings true but is not very descriptive of me in particular. More predictive of my actions is that I try to increase the collective human understanding of the universe. The purpose of my life, hence, is to make others think. I hope I reached my goal.

End Notes

1. The origin of philosophical zombies seems to have gotten lost in the literature, but most relevant articles about it have been collected by David Chalmers and can be found at http://consc.net/online/1.3b, Retrieved March 3rd 2017.
2. Oizumi, M., Albantakis, L., Tononi, G., "From the Phenomenology to the Mechanisms of Consciousness: Integrated Information Theory 3.0," PLoS Comput Biol. 10 (5): e1003588.
3. Tegmark, M., "Consciousness as a State of Matter," Chaos Solitons Fractals 76, 238 (2015) [arXiv:1401.1219 [quant-ph]].
4. Harvey, C. D., Collman, F., Dombeck, D. A., David W. Tank, D. W., "Intracellular dynamics of hippocampal place cells during virtual navigation," Nature 461, 941-946 (15 October 2009).
5. Diester, I., Nieder, A. "Semantic Associations between Signs and Numerical Categories in the Prefrontal Cortex," PLoS Biol 5(11): e294 (2007).
6. With apologies to the more physically inclined readers, I will in the following neglect special relativity and assume there is an absolute time by use of which we can speak of simultaneity so that it's unambiguous what it means for the morphism to be predictive. The definitions I propose here can be made compatible with special relativity by taking into account the finite time it takes information to travel through I/O channels but I space limitations prevent me from elaborating on this. Let me just say it can be done, but isn't relevant for what follows.
7. Liu X, Ward BD, Binder JR, Li S-J, Hudetz AG (2014) Scale-Free Functional Connectivity of the Brain Is Maintained in Anesthetized Healthy Participants but Not in Patients with Unresponsive Wakefulness Syndrome. PLoS ONE 9(3): e92182.
8. Attributed to Michael Russell in: Sean Carroll, "The Big Picture," Dutton (2016), p. 260.

Appendix

A simple way to probe the consciousness-level of an anesthetized patient or a patient with locked-in syndrome might look as follows.

Level one: Testing for level one awareness is straight-forward. Play the patient a sound that repeats in regular intervals, or apply any other regular sensory stimulus, such as touch or, if necessary, direct brain stimulation. If the brain shows any reaction in the same regular sequence as the stimulus that constitutes a simple morphism. (Yes, it's hard to not at least be unconsciously aware of direct brain stimulation.)

Level two: If the patient is able to generate a predictive model, the brain signal should show additional activity if the sequential stimulus—after some period of repetition—changes or suddenly stops. We would be looking, essentially, for a sign of surprise which would demonstrate forecasting ability.

Testing awareness level three and four is much more difficult for it requires finding evidence for the integration of subsystems.

Level three: One way to demonstrate experience, according to our definition, would be to demonstrate that the patient is able to model not only the input signal but the connection between various types of input signals. This would show that the patient is—to some extent—aware of the way their own brain is connected internally and with the environment. This could be done for example by probing whether the patient's reaction to two different kinds of stimulus is the same whether or not the input is synchronized, or where it is coming from.

Level four: Probing cognition isn't all that difficult because we are used to doing it. It can be done by looking for the ability of the patient to learn and react, for example by adding a feedback look to the input signal. A practical way to achieve this may be to continue a signal as long as the patient is displaying a certain brain activity. Over time, the patient should learn of this connection, meaning when the feedback is suddenly discontinued, there should again be a surprise reaction. The presently used tests, such as asking patients to imagine performing a certain action and measuring their response, are much more complex ways to test for level four awareness.

Chapter 9
World Without World:
Observer-Dependent Physics

Dean Rickles

We have found a strange footprint on the shores of the unknown. We have devised profound theories, one after another, to account for its origins. At last, we have succeeded in reconstructing the creature that made the footprint. And lo! It is our own.

Arthur Eddington, 1921.

Physics is to be regarded not so much as the study of something a priori given, but as the development of methods for ordering and surveying human experience.

Niels Bohr, 1961.

The brain is small. The universe is large. In what way, if any, is it, the observed, affected by man, the observer? Is the universe deprived of all meaningful existence in the absence of mind? Is it governed in its structure by the requirement that it gives birth to life and consciousness? Or is man merely an unimportant speck of dust in a remote corner of space? In brief, are life and mind irrelevant to the structure of the universe - or are they central to it?

John Wheeler, 1975.

No element in the description of physics shows itself as closer to primordial than the elementary quantum phenomenon, that is, the elementary device-intermediated act of posing a yes-no physical question and eliciting an answer or, in brief, the elementary act of observer-participancy. Otherwise stated, every physical quantity, every it, derives its ultimate significance from bits, binary yes-or-no indications, a conclusion which we epitomize in the phrase, 'it from bit'.

John Wheeler, 1990.

How can mindless mathematical laws give rise to aims and intention? How can purpose in the universe emerge when there is no purpose at a fundamental level: what do quarks know about it? How is it that the (free) mind can exist in a purely material universe governed by rigid laws of nature? The mind and its goals and purposes are surely subjective while the world is objective. The inanimate world surely has no such thing as goals...

Our question contains two very obvious assumptions: (1) the idea that mathematical laws are indeed mindless; (2) the idea that there are indeed such things in

D. Rickles (✉)
University of Sydney, Sydney, NSW, Australia
e-mail: dean.rickles@sydney.edu.au

© Springer International Publishing AG, part of Springer Nature 2018
A. Aguirre et al. (eds.), *Wandering Towards a Goal*, The Frontiers Collection,
https://doi.org/10.1007/978-3-319-75726-1_9

101

the world as 'aims' and 'intentions'. This essay focuses on assumption (1). There is a long and venerable history, going back at least to Kant, of the view that the regularities and structure we find in the world of physics (laws) correspond to what the mind has itself imposed: we are engaged (to some extent: the degree leading to varying strengths of idealism) in a form of self-study when we study the laws of nature. There have been several subsequent versions of this idea with their roots in physics. What each involves is the view that the laws of physics (and possibly many other features of our scientific representations of the world) are heavily laden with materials from the humans devising such representations and laws. The structure of the universe, on such views, is intimately connected with our own existence.

Our answer to the question, then, is to deny assumption (1): mathematical laws are *not* mindless but are instead infused with features of our cognitive framework.[1] This seems initially shocking. But on closer inspection the shock should be done away with: unless we suppose some kind of mind/matter dualism, we should fully expect the mind to be bound up with worldly things, just as much as worldly things find themselves bound up with the mind. As Schrödinger puts it, to suppose that there is a division between mind and world such that there exists an interaction between them smells of something "magical" or "ghostly" ([5], p. 63). The initial question puts the explanatory cart before the horse. We should be asking how can the mind generate and/or be implicated in laws that appear on the surface to be so mindless.

One can readily speculate about evolutionary accounts involving pattern-finding and the abstraction of invariances from a jumble of data so as to minimize energy and stay alive: it is a very useful thing to be able to predict events that have yet to take place, but we only need to keep track and predict some features of the environment. Indeed, we don't need to be scientists to engage in this kind of behaviour. Even rats do it quite naturally: by training, a rat can *predict* that some stimulus will lead to a reward and so will have discovered a primitive 'law of nature' in its own limited universe. Of course, the question remains: how is it that this is possible in the first place? Ernst Cassirer suggests a way:

> The fleeting, unique observation is more and more forced to the back-ground; only the "typical" experiences are to be retained, such as recur in a permanent manner, and under conditions that can be universally formulated and established. When science undertakes to shape the given and deduce it from definite principles, it must set aside the original relation of *coordination* of all the data of experience, and substitute a relation of superordination and subordination ([2], p. 272).

And later: "We finally call objective those elements of experience, which persist through all change in the here and now, and on which rests the unchangeable character of experience" (ibid., p. 273). Hence, the division into 'objective' and 'subjective' is central to why the question of 'mindless mathematical laws giving rise to mindful behaviour' feels so natural. This division basically bundles a whole bunch of properties thought to be on the side of science (permanence, reality, concrete, physical,

---

[1] As regards assumption (2), I think Carroll ([1], pp. 294–295) rather effectively shows that the notion of 'wants' and 'intention' is really just as *façon de parler*: there are simply situations in which it is more useful than not to describe things in terms of something wanting another thing.

etc.) with 'the objective', and a whole bunch thought to be anti-scientific (spiritual, mental, varying, abstract, etc.) with 'the subjective'.

The problem is, with every scientific experiment there is always a pair of poles, the observer and the observed; the experiencer and the experienced. To cut out one or the other, by saying that laws are only ever 'objective,' leaves an incomplete world. Moreover, the bundling up of the previous properties into objectivity pushes us almost by necessity to feel as if we must eradicate subjects (agents, observers, experiencers) from science and laws. Schrödinger again:

> The scientist subconsciously, almost inadvertently, simplifies his problem of understanding Nature by disregarding or cutting out of the picture to be considered, himself, his own personality, the subject of cognizance. Inadvertently the thinker steps back into the role of an external observer. This facilitates the task [of science] very much. But it leaves gaps, enormous lacunae, leads to paradoxes and antinomies whenever, unaware of this initial renunciation, one tries to find oneself in the picture or to put oneself, one's own thinking and sensing mind, back into the picture. This momentous step—cutting out oneself, stepping back into the position of an observer who has nothing to do with the whole performance—has received other names, making it appear quite harmless, natural, inevitable. It might be called objectivation, looking upon the world as an object. The moment you do that, you have virtually ruled yourself out ([4], pp. 92–93).

Hence, physical reality is deemed tantamount to independence from some arbitrary observer. Inasmuch as one can relate these arbitrary observers to the objective realm, one must 'average over' them in some sense, or take the equivalence class of their perspectives (as one does to get the notion of a 'geometry' from metrics related by diffeomorphisms in general relativity). However, the presence of the observer makes its presence felt at some level since it is still the frames of reference that are being bundled into the equivalence class.

Arthur Eddington famously developed a quasi-idealist worldview in which much of what we might naturally think of as 'the stuff of the world' is 'spiritual' (what we would now call 'mental'). The *physical universe*, as described by the physical sciences and running according to physical laws, is not equivalent to *objective reality*. Instead, Eddington argued for what he called 'subjective realism,' a position that enabled him (in his mind) to deduce the laws of nature and the values of the fundamental constants they involve purely from his armchair (through the study of the basic principles of observation and measurement). In other words, the laws that we often suppose to be entirely mindless and free from any kind of human influence (e.g. subjectivity), are in fact products of the mind:

> An intelligence, unacquainted with our universe, but acquainted with the system of thought by which the human mind interprets to itself the content of its sensory experience, should be able to attain all the knowledge of physics that we have attained by experiment ([2], p. 3).

At this simplistic level, the position sounds rather absurd, and it certainly received its share (often justified) of criticism.[2] To modern ears any denial of the primacy of

---

[2]The main thrust of the criticisms were directed at Eddington's claims to have deduced the fundamental constants (the fine structure constant in particular) by examining the methods of observation rather than the world being observed.

empirical methods for learning about the construction of the world is met with an incredulous stare—it precisely contravenes the objectivity mentioned earlier. However, it is important to note that this is not idealism in the orthodox sense of, e.g., Bishop Berkeley: there are *particular facts* about the world too, and these must be derived from experience.[3] More importantly, the subjectivism is not taken too far: there are overlapping aspects of the world in multiple subject's consciousness that are grounded in an external world that itself is not part of the content of any observer's mind. Facts about the distance to Mars cannot be deduced from pure reason. But claims about possibility and necessity (e.g. of what processes can and cannot occur) are the stuff of minds.

Eddington had his own criticism of current science which he thought was far more 'mystical' than his own ideas. For example, unlike in earlier times where clockwork and engineering with the dominant paradigms for scientific credibility ('the tyranny of the engineer'), modern physics is beholden to more abstract ideals ('the tyranny of the mathematician'). It is possible that this use of mathematics imposes 'blinkers' on the view of the world more so than those earlier physicalistic models. The success of the mathematical approach according to Eddington stems primarily from the fact that this approach gets to set the terms of its own success: a numerical test.[4] There's objectivity in numbers for sure, but the kinds of mathematics we choose to use is still a reflection of features of observers.

If this stretches incredulity too far, let us consider a modern version of what is in many respects a similar position: Wheeler's 'It From Bit' and the related notion of existence as a 'self-excited circuit'. Whereas Eddington was inspired to his more subjectivist physics by general relativity, Wheeler was pushed there by quantum mechanics, especially his own delayed-choice experiment which shows that experimenters (agents) can decide, after the event, whether a photon was in two places or one in the context of a double-slit type experiment.[5]

> 'It from bit' symbolizes the idea that every item of the physical world has at bottom—at a very deep bottom, in most instances—an immaterial source and explanation; that what we call reality arises in the last analysis from the posing of yes-no questions and the registering of equipment-evoked responses; in short, that all things physical are information-theoretic in origin and this is a participatory universe ([6], p. 5).

---

[3]The position he espoused is known as 'Selective subjectivism'. As he puts it, this "the modern scientific philosophy, has little affinity with Berkeleian subjectivism, which, if I understand it correctly, denies all objectivity to the external world. In our view the physical universe is neither wholly subjective nor wholly objective—nor a simple mixture of subjective and objective entities or attributes" (*The Philosophy of Physical Science*, 1938, Cambridge University Press, p. 27).

[4]Eddington was pushed to this viewpoint by the changes brought about by general relativity. His final work, *Fundamental Theory*, explicitly distinguished between observables and unobservables, with the former being those quantities that can be ascertained by a measurement procedure, and the latter not thus capable because they contain mathematical baggage (they include an auxiliary mathematical component that cannot be observed).

[5]Wheeler called himself a "radical conservative'. It's no surprise that Wheeler's longterm student and friend Peter Putnam was obsessively interested in Eddington's philosophical scheme. I don't doubt that Putnam influenced Wheeler's it from bit through his discussions of Eddington.

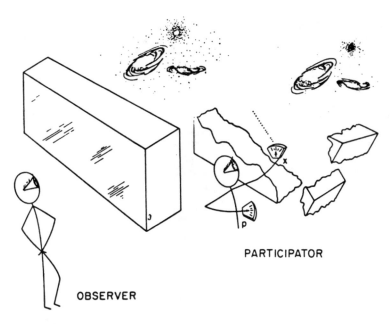

PARTICIPATOR

OBSERVER

Fig. 9.1 A Wheelerian participant: active in the universe, unlike a classical observer. *Source* J. Wheeler, "Beyond the Black Hole", in H. Wolf. (ed.), *Some Strangeness in Proportion*, 1980, Reading, MA: Addison-Wesley, p. 355

The observer participates in the defining of reality (see Fig. 9.1). In many ways this inclusion of the observer as an active player in the development of the universe is somewhat like the inclusion of spacetime geometry. It extends background independence.

Wheeler takes this observer involvement to extremes as can be seen in another neat little picture (a 'self-excited circuit', Fig. 9.2). This picture refers to the fact that, in quantum theory at least, our observations determine the very reality we are studying (by choosing which experimental questions to put to nature), so that we are in effect studying aspects ourselves when we examine the quantum world. This can even be extended to observations way into our own past (before there were observers: *very* delayed choice experiments). In this case, we bring about our universe, lifting it into existence by its bootstraps. Not only are laws not mindless, minds (or observers) are at the root of their very existence and the arena in which they operate.

Like Eddington's selective subjectivism, this is an epistemological theory that explicitly incorporates 'observer-selection effects' (the idea that in some sense our presence as observers influences *what* we observe). It is thus deeply anthropic. But it is also intended to be ontological: what there is (the world itself) is built up from the specific yes/no questions observers put to Nature. What *is* is what *happened*.

Fig. 9.2 The self-excited
circuit of John Wheeler,
representing the idea that the
universe 'bootstraps' itself
into existence by observing
itself, thus creating
'phenomena'. *Source* J.
Wheeler, "Beyond the Black
Hole", in H. Wolf. (ed.),
*Some Strangeness in
Proportion*, 1980, Reading,
MA: Addison-Wesley, p. 362

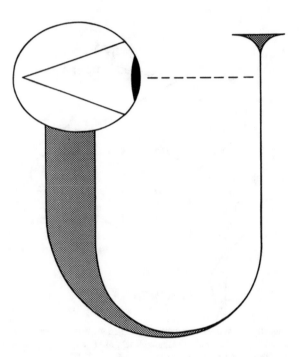

And what happened is guided to a certain extent by the experimenter.[6] This general
participatory scheme has been fleshed out in the form of a novel interpretation of
quantum mechanics: QBism.

> QBism says that when an agent reaches out and touches a quantum system—when he per-
> forms a quantum measurement—that process gives rise to birth in a nearly literal sense. With
> the action of the agent upon the system, something new comes into the world that wasn't
> there previously: It is the "outcome," the unpredictable consequence for the very agent who
> took the action ([3], p. 8).

As the architects of QBism make clear, Wheelerian participatory principles are at
the root of this approach. It introduces subjectivist elements by simply treating probability via a Bayesian framework rather than some objectivist framework.

---

[6]Wheeler envisages a scaled-up version involving photons having travelled a billion light years
from a quasar, separated by a grating of two galaxies (to act as lenses offering two possible paths
for the light), to be detected at the Earth using a half-silvered mirror at which the twin beams can be
made to interfere. For Wheeler, this means that the act of measurement (our free choice) determines
the history of that entire system: actions by us NOW determine past history THEN (even billions of
years ago, back to the earliest detectable phenomena, so long as we can have them exhibit quantum
interference). It is from this kind of generalization of the delayed-choice experiment that his notion
of the Universe as a self-excited circuit comes: the Universes very existence as a concrete process
with well-defined properties is determined by measurement. Measurement here is understood as
the elicitation of answers to 'Yes/No' questions (e.g. did the photons travel along path A or B?):
bit-generation (gathering answers to the yes/no questions) determines it-generation (the universe
and everything in it). However, Wheelers notion does not privilege human observers, but rather
simply refers to an irreversible process taking uncertainty to certainty.

The orthodoxy amongst scientists (especially physicists) is that the universe is a purposeless block. It just *is*. The goal of the scientist is to uncover its invariant features, its laws. But where does the scientist's goal come from if the very universe they are studying, in which they are also embedded, is supposed to possess nothing of the sort? The scientist will quickly respond that it (their very own goal) is merely illusory: their aims and dreams are simply the stuff of physics too, and as such are determined by the very same laws they study. Any choice the scientist might appear to make was in accord with the laws of nature and was determined by them. These observer-dependent (or, less strongly: observer-inclusive) approaches do not involve a 'ready-made world.' In a letter to Gödel from 1974 Wheeler wrote, in apparent consternation, that he had just learned at a party that Gödel "believes in the existence of what is sometimes called 'an objective universe out there'. Whether through general relativity or quantum theory, our best theories allow interpretations that put observation center stage.

Let us end with a few selections from Wheeler's final blackboard:

- 8. Physics has to give up its impossible ideal of a proud unbending immutability and adopt the more modest mutability of its sister sciences, biology and geology.
- 9. If the kingdom of life and the highest mountain ranges are brought into being by the accumulation of multitudes of small individual processes, it is difficult to see what else can give rise to the universe itself.
- 10. What other possibility is there for "law without law" except the statistics of large numbers of lawless events?
- 11. No elementary process is as attractive for this statistics as the elementary act of observer-participatorship.
- 15. No working picture that can be offered today is so attractive as this: the universe brought into being by acts of observer-participatorship; the observer-participator brought into being by the universe ("self-excited circuit").

And so we see again the idea that the world is in some sense mind-stuff (world without world), or at least infused with some kind of mind stuff, through observations and so on. The viewpoint expressed in this essay is that a pressing problem of physics is to recognize that our role as observers is more deeply embedded in our theories and laws than is often realised. Whether we wish to go to the extremes of Eddington and Wheeler is another matter...

Acknowledgements I would like to acknowledge the Australian Research Council for funding via a Future Fellowship: FT130100466.

References

1. Carroll, S.: The Big Picture. Dutton (2016)
2. Cassirer, E.: Substance and Function. Dover (1923)
3. Fuchs, C.: QBism, the Perimeter of Quantum Bayesianism (2010). https://arxiv.org/pdf/1003.5209.pdf

4. Schrödinger, E.: Nature and the Greeks. Cambridge University Press, Cambridge (1954)
5. Schrödinger, E.: My View of the World. Cambridge University Press, Cambridge (1964)
6. Wheeler, J.A.: Information, physics, quantum: the search for links. In: Zurek, W. (ed.) Complexity, Entropy and the Physics of Information. Westview Press (1990)

Chapter 10
The Role of the Observer in Goal-Directed Behavior

Inés Samengo

A bunch of nucleic acids swim among many other organic compounds forming a cytoplasmatic soup, and somehow, manage to arrange themselves into precisely the sequence required for DNA replication. Carbon dioxide molecules steadily stick to one another materializing a solid tree trunk out of a tiny seed. Owls eat the young bats with poor navigation ability, thereby improving the eco-location proficiency of the species. The neurons in a dog's brain fire precisely in the required sequence to have the dog bury its bone, hiding it from other dogs. The wheels, break, and clutch of a self-driving car coordinate their actions in order to reach the parking area of a soccer field, no matter the initial location of the car, nor the traffic along the way. The limbs of the Argentine soccer players display a complex pattern of movements that carry the ball, through kicks and headers, at Messi's feet in front of the keeper... kick... **goal**! This essay is about goals. In all the above examples, a collection of basic elements, following local and apparently purpose-less laws, manage to steer the value of certain variables into some desired regime. The initial state is rather arbitrary, and yet, the agents manage to adaptively select, out of many possible actions, the maneuvers that are suited to conduct the system to the desired goal. Throughout these seemingly intelligent choices, order appears to raise from disorder. Scattered nucleotides become DNA. Air and dust become trees. Owl hunger becomes sophisticated eco-location organs. Neural activity becomes a buried bone. A car anywhere in the city becomes a car at a specific location. A football anywhere in the stadium becomes a football in the goal. How do the components of each system know what to do, and what not to do, in order to reach the goal? This is the question that will entertain us here.

I. Samengo (✉)
Department of Medical Physics and Instituto Balseiro,
Comisión Nacional de Energía Atómica, San Carlos de Bariloche, Río Negro, Argentina
e-mail: ines.samengo@gmail.com

© Springer International Publishing AG, part of Springer Nature 2018   109
A. Aguirre et al. (eds.), *Wandering Towards a Goal*, The Frontiers Collection,
https://doi.org/10.1007/978-3-319-75726-1_10

One crucial characteristic of goal-directed agents is that they are flexible: They reach the goal from multiple initial conditions, and are able to circumvent obstacles. For example, in DNA replication, the initial state is one out of many configurations in which nucleotides can be spatially distributed in a solution of organic compounds. The final state, the goal, is the precise spatial arrangement of those same nucleotides within the newly constructed DNA strand. In the soccer stadium, the ball may be initially in any location, the final state is the ball at the goal. Multiple initial states are hence mapped onto a single final state, as in Fig. 10.1. In physical terms, the non-injective nature of this mapping implies a reduction in entropy. Here, entropy is understood as the quantity obtained by calculating $S = -\sum_i p_i \log p_i$, where the sum ranges over all the states of the system, and p_i is the corresponding probability [16].

Admittedly, the final state need not be strictly unique. In DNA replication, permutations of equal nucleotides are still allowed in the final state, and occasionally, there might also be a few errors in the replication process. Dogs may consider more than a single location for the concealed bounty, and Messi may choose to shoot the ball anywhere inside the 24 ft wide by 8 ft high of the goal. The restricted amounts of freedom in the final state, or even the occasional failures to reach the desired objective in individual trials (shooting an own goal, for example), does not compensate the reduction of entropy that takes place in multiple trials, not at least if the system is to be understood as steered towards a goal. If entropy does not decrease, evolution is either a few-to-many random mapping, or a one-to-one deterministic rule, but not a many-to-few directed process. Goals can only be arrogated to the latter, because only the latter are based on versatile strategies that craftily avoid obstacles to target the objective.

In physics, we often associate entropy increments with information losses, and entropy reductions with information gains. Here I am taking the opposite view: Entropy reductions are associated with information losses. The two views are not incompatible, they simply refer to different points of view. When physicists claim that information is lost because entropy increases, they are typically dealing with a

Fig. 10.1 In DNA replication, many initial states are mapped onto a single final state. The entropy is therefore high at the beginning and low at the end

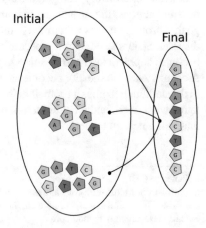

closed system described by macroscopic variables. The final macroscopic state does not allow one to deduce the initial macroscopic state, since the mapping between the two is non-injective. Were we to know the detailed final micro-state, however, we would be able to deduce the initial micro-state, due to the reversibility of physical laws. This ability, however, is lost when only keeping the coarse scale. In this essay, taking a different point of view, I associate entropy reductions with information losses. When a goal-oriented system evolves in the direction of decreasing entropy, the final microscopic state does not allow one to deduce the initial microscopic state, even if evolution is governed by reversible physical laws. The clue lies in the fact that goal-oriented systems are open, and they interact with degrees of freedom we are not keeping track of.

So far I have argued that goal-oriented behavior always brings about an entropy reduction. I now want to demonstrate the reciprocal statement: If a system reduces its entropy, a goal can be ascribed to the process. Therefore, entropy reduction and goal-oriented behavior are in a one-to-one correspondence. In an entropy-reducing system, a goal can always be defined by the restricted set of values that the variables acquire in the final state: the target DNA sequence, the buried bone, the ball at the goal. Of course, the reduction in entropy must first be verified: multiple trials tested with a broad set of initial states must evolve into a small final set. A car that in a single trial travels from one location to another is not guaranteed to be a self-driving car. Only if the initial location has proven to be arbitrary, and the traffic conditions variable, can goal-directed behavior be arrogated.

The notion of entropy is subtle, since its value not only characterizes the state of a physical system, but also, the way it is described. When the universe is described at its utmost basic level (assuming there is one such level), all we have is a collection of fundamental particles evolving from some initial state, classical or quantum. If the state of all particles is specified, the total entropy of the universe vanishes. Time reversibility of the laws of physics dictates entropy to remain zero for all past and future times. Therefore, there is no way to attain neither an increase nor a reduction in entropy. Energy dissipation and goal-directed behavior, hence, are absent from the complete description. We need to blur our point of view to give them a chance, either by restricting the description to macroscopic variables, or to subsystems. In fact, the main conclusion of this essay is that an observer with a very special point of view is required for agency to exist.

If the information about the initial conditions is apparently lost in goal-oriented behavior, then such information must be somehow concealed in degrees of freedom we are not keeping track of. It may be carried by variables that are too microscopic to be monitored, or by fluid degrees of freedom that, by the time the goal is reached, have already exited the subsystem under study. What we track, and what we ignore, hence, plays a crucial role in agency [13].

To be consistent with the second law of thermodynamics, processes where entropy decreases are only possible in open systems that somehow interact with the external world. Originally, they were supposed to require an energy influx. This is, however, not a necessary condition: Sometimes, the sole exchange of information suffices. A good example is Maxwell's Demon [11]. Suppose we have a gas enclosed in

Fig. 10.2 A demon controls the sliding door, allowing particles to pass from right to left, but not the other way round. The initial state of every molecule that the demon has already acted on (letting it pass or not) is recorded in its memory, and depicted in blue

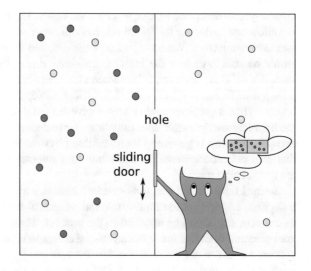

two adjacent chambers communicated by a small hole in the wall between them (Fig. 10.2). The hole may or may not be covered by a sliding door controlled by a demon. Initially, both chambers have equal pressure and density. The demon then opens or closes the hole selectively, depending on whether a molecule approaches from one side, or the other. Molecules coming from the right are allowed to pass into the left chamber, but not the other way round. As time goes by, molecules accumulate on the left side, eventually leaving the right side empty. The collection of all gas molecules can be interpreted as performing goal-directed behavior: No matter the initial state, gas is gradually compressed into the left chamber. This final state can be conceived as a goal, and it comprises a reduction in entropy: initially each particle can be anywhere in the two chambers, and in the final state, they are all in the left side. Arrogating purpose, in this case, is to assume that the gas—who takes the role of the agent—*wants* to shrink. Other verbs may be used (*tends to, is inclined to*, etc.), but the phrasing is irrelevant. As uncanny as it may seem, arrogating purpose to the gas is a rather accurate description of the gas' phenomenology.

The gas + demon is a toy model of a closed system, so no interaction with the outside world is allowed. To perform the task, the demon needs to acquire information on the location of each molecule approaching the hole, to then decide whether to let it pass or not. In a slightly modified version of this system, Bennett [1] demonstrated that the storage of information in the demon's memory can be done with no energy expenditure, as long as the memory is initially blank, and there is plenty of storage capacity. The work required to move and stop the door, as well as the energy needed to measure the position of particles and to maintain the demon alive, can also be made as small as desired, simply diminishing mechanical friction, and moving slowly. The demon is however not allowed to delete the acquired information, because information erasure requires energy consumption, at a minimal cost of $k_B T$ per erased bit [9]. Therefore, as time goes by, the information of the initial location of each gas molecule is erased from the gas, and copied onto the demon's memory.

The gas gradually reduces its entropy only if we are careful to exclude the demon from what we define as *the system*. If we include the demon (and its memory), entropy simply remains constant, since all the details of the initial state are still stored. Depending on the observer's choices, then, entropy may or may not decrease, meaning that arrogating agency may or may not be possible.

A subsystem can only decrease its entropy if it somehow gets rid of initial conditions. In DNA replication, after the addition of each new nucleotide to the developing strand, the initial location of the free nucleotide determines the final configuration of the mediating enzymes. The information of the initial spatial arrangement of nucleotides is thereby transferred to the 3-dimensional configuration of nearby proteins. If enzymes are not restored to their functional state, the process cannot be iterated. So enzymes, in turn, must pass the information on somewhere else. This transfer is actually the important point in the emergence of goal-directed behavior. Energy consumption is only helpful if energy is *degraded* in the process: ordered energy sources must be transformed into disordered products. In animal cells, order arrives as glucose and oxygen molecules. Disorder exits as carbon dioxide, water and faster molecular motion (heat). The input degrees of freedom, specifically in the case of glucose, are conformed of atoms tidily organized into large molecules. The output degrees of freedom are transported by smaller molecules, amenable to be arranged in many more configurations.

The laws of physics are ultimately reversible, so initial conditions cannot be truly erased, they can only be shuffled around. As an example, Edward Fredkin studied how non-dissipative systems, such as our universe, may perform the usual logical computations (AND, OR, etc.), which are themselves non-invertible [6]. We know that 0 AND 1 = 0. However, knowing that the result of the operation is 0 does not suffice to identify the two input variables. If the computation is performed by an ultimately non-dissipative system, the information of the initial input variables must be somehow moved into some other variable, albeit perhaps not in a manner that is easily accessible. Fredkin's solution was to prove that computing required some extra input variables, not needed for the computation per se, but mandatory for the information balance. When performing a single logical operation (say, for example, AND), the additional variables are in a well defined state (no uncertainty), and throughout the computation, they acquire so-called garbage values (garbage because they are not required to perform the computation), that represent those input degrees of freedom that cannot be deduced from the output. Copying part of the input into garbage variables at the output ensures that no information is lost, and the computation becomes feasible in a non-dissipative physical substrate.

Ascribing agency is all about ignoring who really did the job (the Universe, to put it grandly), and arrogating intentionality to an entropy-reducing subsystem. The task of the observer is to design the borders of the subsystem so as to allow ordered degrees of freedom to be progressively incorporated, and/or disordered ones to be eliminated. If the goal is to be achieved repeatedly, a steady flow of order is required,

Fig. 10.3 Goal-directed
systems (blue ball) eat up
ordered degrees of freedom,
and produce disordered
degrees of freedom

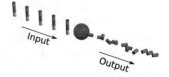

as well as a regular garbage collection service. Purposeful agents, hence, only emerge from sub systems that eat up order (Fig. 10.3). Broadly speaking, they can be said to breathe, or to be endowed with metabolism, even if they need not be alive in the biological sense.

Maxwell's demon hid the initial conditions of the gas in its memory. The dog, the self-driving car, and the soccer players, all hide their own initial state and that of the environment inside their memories. Memories can of course be erased, but erasures consume energy, and they are ultimately no more than flushing initial conditions into the high-entropy products of energy degradation. For a long time, scientists failed to include memories as part of the systems under study, so goal-directed behavior sometimes appeared paradoxical. Here I argue that observers attribute agency by disregarding initial conditions. Of course, observers are free to delineate the borders of the subsystem under study as they wish. They can always shape the limits of what they define as *the agent* in such a way as to have it do all sorts of wonderful things, as achieve goals, and reduce entropy. The agent must be fed with order, and the mess must be cleaned up, but still, it can be done. The natural question is therefore: What is interesting in goal-directed behavior if the observer is allowed to engineer the very definition of the agent, in order to get the desired result? The main conclusion of this essay is that the interesting part is the observer, who decides what to describe and what to ignore. Plants grow because what we define as a plant is the stuff that grows every spring, and not the dirt left on the ground every autumn. Species improve because we restrict the definition of a species to the material that a posteriori is seen as successful, and exclude the corpses left behind of those who failed. Cell division seems to be a productive business because the waste products are not defined to be part of cells. Returning to the question posed above "How do the components of the system know what to do, and what not to do, in order to reach the goal?", we can now provide an answer. Components know nothing, observers do. Just as photographers select an arbitrary plane in the visual world where to focus an image and engender a sharp object (Fig. 10.4), so do observers choose which variables compose the system, and which do not, so that a goal emerges.

The point of view of the observer may have striking consequences. A system that reduces entropy can be used to produce work. Once Maxwell's demon has collected all particles in the left chamber, the pressure that they make on the middle wall can be used to move a piston. This is an example of an information-to-energy conversion. Initially, the demon's memory was empty, so it was ordered. As time goes by, the demon's memory becomes full (disordered), and the gas becomes ordered. If the ordered gas is used to move a piston, then the initial order in the demon's memory has been converted to physical energy. Therefore, work can be extracted from order

Fig. 10.4 Observers, just as photographers, selectively focus on an aspect of reality, to satisfy their cognitive appetite. Potograph kindly supplied by Lucía Samengo

inasmuch as it can be extracted from a potential, or from heat. These transformations are ruled by the laws of thermodynamics of information [14], and have been verified experimentally [18]. In a closed system, however, conversions cannot be iterated indefinitely, since they always involve the production of garbage variables, that can only be restored to their ordered initial state with the expenditure of an amount of energy that is at least as large as the one obtained through the produced work. In other words, if the demon's memory is not wiped, the poor devil's head turns boiling hot, and useless. This effect is also evident in students and workaholics around the globe, providing a performance-based justification of holidays and crappy TV shows.

Should we be amazed that the world we live in allows observers to create agents? Could we not live in a universe where assigning agency were downright impossible? I would be very much surprised if it were so. The impossibility to define goal-directed behavior would mean that no subsystems exist where entropy decreases. The global entropy growth that takes place in the whole universe should develop uniformly and monotonously, allowing for no local departures from the global trend. That would be ordered indeed! I do not expect disorder to arise in such an orderly manner. In fact, a branch of physics is now devoted to determining the conditions in which spontaneous development of order is to be expected [8, 15].

In the picture presented so far, all the interesting events seem to take place in the observer's creative act. Any decrement of entropy, no matter how trivial, appears to suffice for an observer to ascribe agency. We are demanding little of the world, and a lot of the observer. But does the evolving world not have organizational merits of its own? Throughout the eons, profuse RNA replication in free solutions gave rise to prokaryote cells, who in turn evolved into eukaryotes, from which multi-cellular organisms appeared, all the way up to the ever growing branches of the tree of life. In the way, conscious humans, civilization, and artificial intelligence emerged. As well as a lot of garbage, as environmentalists wisely remind us.

In this spiralling progression, evolution seems to be striving towards a runaway escalation of sophistication and design [5]. In particular, agents seem to become increasingly selective. Compare the strategy of a replicating DNA molecule in a bacterium with the one of Menelaus of Sparta to recover Helen of Troy, and thereby, ultimately manage bisexual reproduction. Both construct a new DNA strand, but their strategies are not equally elaborate. The efforts, the tactical planning, and the mess left on the ground by Menelaus are hardly comparable to those of the bacterium. Highly evolved agents pursue goals that are clearly more complex than those their distant forefathers—for the sake of the argument, I am here conceiving Menelaus as an evolved bacterium. Is the development of refined tactics something that only depends on the observer's creativity, or is it something actually taking place in the world, independently of observers? In the end, evolution seems to be an objective process, governed by laws and principles that are not related to whom may observe them.

The fact that observers play a protagonic role in goal-oriented behavior does not mean that they are the only characters in play. Observers are indispensible to produce agents. In the absence of agents, no subsystems are cut out of the wholeness of the cosmos, and complexity cannot be measured. So in this respect, the evolving world and the increase in its complexity are the result of the act of observation. Yet, observers are not wizards that cook up their images out of thin air. Their involvement extends as far as to decide what to observe and what to ignore, but not beyond. The fact that observers spot goal-oriented agents within the unity of reality does not preclude the observed agents to have a dynamics of their own, a dynamics not controlled by observers, and which on our planet, happened to generate order and intelligence through evolution. Starting with Darwin, many have studied the laws governing the emergence of order, and I am not here denying those laws. My point is that during evolutionary development, the increasing sophistication is evenly balanced by an equally large amount of garbage. Only biased observers focus on the constructive side alone, concluding that evolution is striving towards perfection. And only biased observers focus on the destructive side alone, concluding that disorder inevitably grows. Increases or decreases in entropy are a matter of point of view. At least, inasmuch as we hold to reversible physical laws.

In fact, the very gift that observers possess, the ability to spot intentionality, can be itself understood as the result of evolution. If we look at the world with unbiased eyes, we see many things changing, some relevant and some irrelevant to our fate. To have a chance to survive and pass on our genes, it is important for us to identify subsystems from which predictions can be made, focusing particularly on those prediction that modify our fitness. As a consequence, brain-guided observers ascribe agency and make predictions. They do so because the role of a brain is to model the world around its carrier, so that effective survival strategies can be implemented. Mental models must capture the regularities of the world, and discard the noise. Here, noise is defined as the degrees of freedom that are irrelevant to predicting those features of the environment that affect the observer's fitness. It would be a waste of resources, if not impossible, for us to represent in our brains all what happens in a dog's brain. Much more efficient is to ascribe agency, and conclude that the dog *wants* to bury the bone. We cannot follow the evolution of all the bats that were eaten by owls, we therefore conclude that the predation of owls *sharpens* the eco-location capacity of bats. We do not care for the details with which self-driving cars are programmed, we just think of them as goal-directed. We need an economic description, so we assign agency [3].

Reduced models make predictions in terms of heuristic laws that are naturally less accurate than the completely detailed description. The laws of psychology, to put an example, are not as reliable as the laws of chemistry. Yet, given the limited resources of the observer, the arrogation of agency is convenient if predictions can be made, even if somehow inaccurate, at a sufficiently low computational cost. Actually, observers do not assign agency to all the entropy-reducing systems they meet. Purpose is only arrogated to subsystems for which there is no evident source of order, or for sources that are too costly to represent. The balance between costs and benefits depends on each case. If we only look at the gas controlled by Maxwell's demon, ignoring the demon itself, we conclude that the gas wants to shrink. If, however, the demon takes weekends off, the purposeful model of the gas loses accuracy. A more expensive representation discerning between week days and weekends is needed. Assigning agency may or may not be a convenient strategy, depending on the trade-off between the economy of the representation and the prediction errors it induces. Arrogating agency in excess, for example by believing that all what happens is maneuvered by some obscure intentionality, yields a poor prediction strategy.

Observation is the result of development: Observers learn how to observe, and they do so within the framework of learning theory [10]. They are first exposed to multiple examples of the process, that act as the training set. Before learning, the final state can only be predicted from the initial one if all degrees of freedom are tracked—a representation capacity that observers typically lack. Making the best use of their resources, observers explore the power set of the system (the set of all subsets of the system) and search for some entropy-reducing subset from which an agent and a goal can be defined. They then discard the superfluous degrees of freedom, thereby compressing information. Yet, if the subsystem does indeed reduce entropy, they are still able to make predictions. With successful predictions the world begins to makes sense, so ascribing agency is in a way equivalent to constructing knowledge. In fact, the construction of knowledge can be argued to be the essence of a mind.

In the last paragraphs, we have been observing observers. In doing so, we have climbed one step higher on the hierarchical ladder of observation. We have concluded that purposeful agents do not exist per se, they are a mental construct of observers. This view may be easily accepted when regarding the agents around us, but becomes more problematic when it applies to ourselves. We feel our own desires, and experience a vivid sensation of accomplishing our objectives out of our own free will. The idea that we exist as agents and we reach our goals only because some alien observer defined our boundaries, sounds outlandish at the very least. In this context, we face two alternatives. Either we try to understand ourselves as agents in the same terms we conceive other agents, human or not, or we believe our own individual agency is a unique process that follows different rules from those of all other agents around us. What follows is an effort to ascribe to the first option.

The difficulty arises because it is not easy to conceive ourselves as simultaneously the subject and the object of observation. The usual understanding of the act of observation implies the existence of a subject and an object that are different. And the usual understanding of the concept of self implies a unit entity, not divisible in subparts. And yet, within a physicalist's point of view, the observer that creates the agent in us cannot be situated in any other place than in the same brain where the self emerges. Whether the observer coincides with the self, whether it only partially overlaps with it, whether it contains it, or is contained by it, I do not dare to assert. I conjecture, however, that each brain creates its own self following the same principles with which it ascribes agency to external factors. We can much better predict the movements of our own limbs than the movements of distant objects. So it makes sense to define ourselves as an agent whose spatial limits coincide with the surface where predictability drops: our skin. We can also predict some of our thoughts, and sensations, and such thoughts and sensations are intimately related with the predictions of the state of our body. It therefore makes sense to include our thoughts in the notion of self. Moreover, past events are a crucial ingredient in the ability to predict the future, so our memory is also included as a part of our self. In this view, our self is the set of highly connected processes from which intimate predictions can be made.

One huge difference between creating external agents and creating our own self is that the latter involves a vastly larger number of degrees of freedom. Those degrees of freedom, moreover, are typically only accessible to the local subsystem. They include the mental processes to which we have conscious access, encompassing external sensory input, and the detailed state of our body. The latter has been proposed as the base for emotion, and the higher-level neural patterns triggered by such a state, the base of feeling [2].

Douglas Hofstadter has suggested that the circular nature of the mind observing itself is the essence of the self [7]. I am not sure, however, whether the strangeness of this loop constitutes an actual explanation of the self, or simply a way to bind the two loose ends together and worry no more. It could also be the case that what we perceive as a unitary self is in fact a whole collection of disperse mental processes [12], inside which multiple observers coexist, although separately unaccessible. In the end, consciousness has been equated with complex and indivisible information processing [17], so accessing subprocesses may not be possible. Dennett however very strongly argues that if such mental subprocesses can be considered multiple observers, there is no such thing as a hierarchy, and even less, an ultimate observer [4].

I am afraid I am unable to provide a finished picture of agency when going all the way up to the self. I hope, however, to have built a sensible image of other less intimate agents. The main conclusion of this essay is that the interesting part of agency is the observer. Physics does not make sense, observers make sense of it. Life does not have a meaning, we give it a meaning. Life may not even be fundamentally different from non-life, it may just be a collection of subsystems that appear to have goals. Goal-directed behavior does not exist if we do not define our variables in such a way as to bring goals into existence. Bringing goals into existence is a task that brains perform naturally, because they have evolved to model the world and predict the future. One fundamental agent that has emerged inside each one of us is the self. Although the mechanisms behind the emergence remain unclear, the uprise of a self might have produced an evolutionary advantage. I conjecture the self may have arisen as a compressed representation of intimate events, from which accurate predictions can be made. By distinguishing ourselves as a special part of the cosmos we produce meaning. We acquire a sense of self-preservation that gives rise to a sense of identity. The self is a point of view from which to enunciate even the most basic statements, all the way up from *cogito ergo sum.*

References

1. Bennett, C.H.: The thermodynamics of computation-a review. Int. J. Theoret. Phys. **21**(12), 905–940 (1982)
2. Damasio, A.R.: The Feeling of What Happens: Body and Emotion in the Making of Consciousness. Harcourt Brace & Co, San Diego (1999)
3. Dennet, D.C.: The Intentional Stance. MIT Press, Cambridge (1989)
4. Dennet, D.C.: Consciousness explained. Little, Brown and Co, Boston (1991)
5. Dennet, D.C.: Darwin's dangerous idea: evolution and the meaning of life. Simon & Schuster, New York (1995)
6. Fredkin, E., Toffoli, T.: Conservative logic. Int. J. Theoret. Phys. **21**(3–4), 219–253 (1982)
7. Hofstadter, D.: I Am a Strange Loop. Basic Books, New York (2007)
8. Kauffman, S.A.: The Origins of Order: Self Organization and Selection in Evolution. Oxford University Press, New York (1993)
9. Landauer, R.W.: Irreversibility and heat generation in the computing process IBM. J. Res. Dev. **5**(3), 183–191 (1961)
10. MacKay, D.: Information Theory, Inference and Learning Algorithms. Cambridge University Press, New York (2003)
11. Maxwell, J.C.: Theory of heat. Longmans, Green, London, New York (1908)
12. Minsky, M.: Society of Mind. Simon and Schuster, New York (1986)
13. Norretranders, T.: The User Illusion: Cutting Consciousness Down to Size. Viking, New York (1999)
14. Parrondo, J.M.R., Horowitz, J.M., Sagawa, T.: Thermodynamics of information. Nat. Phys. **11**, 131–139 (2015)
15. Perunov, N., Marsland, R.A., England, J.L.: Statistical physics of adaptation. Phys. Rev. X **6**(2), 021036 (2016). 12 pages
16. Shannon, C.E.: A mathematical theory of communication. Bell Syst. Tech. J. **27**(3), 379–423 (1948)
17. Tononi, G.: Consciousness as integrated information: a provisional manifesto. Biol. Bull. **215**, 216–242 (2008)
18. Toyabe, S., Sagawa, T., Ueda, M., Muneyuki, E., Sano, M.: Experimental demonstration of information-to-energy conversion and validation of the generalized Jarzynski equality. Nat. Phys. **6**, 988–992 (2010)

Chapter 11
Wandering Towards Physics: Participatory Realism and the Co-Emergence of Lawfulness

Marc Séguin

> No way is evident how physics can bottom out in a smallest object or most basic field or continue on to forever greater depths (…) [the] possibility presents itself that the observer himself closes up full circle the links of interdependence between the successive levels of structure.
>
> John Archibald Wheeler, *Genesis and Observership*[2]

In his 1979 essay "Frontiers of Time",[3] John Archibald Wheeler imagined a peculiar version of the game of twenty questions:

> About the game of twenty questions. You recall how it goes—one of the after-dinner party sent out of the living room, the others agreeing on a word, the one fated to be questioner returning and starting his questions. "Is it a living object?" "No." "Is it here on earth?" "Yes." So the questions go from respondent to respondent around the room until at length the word emerges: victory if in twenty tries or less; otherwise, defeat. Then comes the moment when we are fourth to be sent from the room. We are locked out unbelievably long. On finally being readmitted, we find a smile on everyone's face, sign of a joke or a plot. We innocently start our questions. At first the answers come quickly. Then each question begins to take longer in the answering—strange, when the answer itself is only a simple "yes" or "no". At length, feeling hot on the trail, we ask, "Is the word 'cloud'?" "Yes," comes the reply, and everyone bursts out laughing. When we were out of the room, they explain, they had agreed not to agree in advance on any word at all. Each one around the circle could respond "yes" or "no" as he pleased to whatever question we put to him. But however he replied he had to have *a* word in mind compatible with his own reply—and with all the replies that went before. No wonder some of those decisions between "yes" and "no" proved so hard!

In the regular version of the game, some word is selected before the questioner starts to ask questions. But in this version, the final word "emerges" from the interplay of all the participants. Of course, Wheeler envisioned this story as an allegory for the strange world of the quantum. As he goes on to explain,

M. Séguin (✉)
Collège de Maisonneuve, Montréal, QC, Canada
e-mail: mseguin@cmaisonneuve.qc.ca

© Springer International Publishing AG, part of Springer Nature 2018
A. Aguirre et al. (eds.), *Wandering Towards a Goal*, The Frontiers Collection,
https://doi.org/10.1007/978-3-319-75726-1_11

> There was a "rule of the game" that required of every participant that his choice of yes
> or no should be compatible with some word. Similarly, there is a consistency about the
> observations made in physics. One person must be able to tell another in plain language
> what he finds and the second person must be able to verify the observation.

Can we read in this story even more than Wheeler intended, and suppose that the
physical world itself emerges from the interplay of the participants in the "game of
life"? It doesn't seem possible: if the participants owe their existence to the physical
world that they inhabit, they cannot exist prior to it and cannot bring it into existence…
unless one allows for a *strange loop*, like in those Esher drawings where two hands
mutually draw each other, or where an ever-ascending staircase arranged in a loop
comes back to its starting height. Strange loops are fun to contemplate in art and
in playful philosophy, but surely, one cannot seriously consider using the idea as a
solution to the riddle of existence?

And yet, the alternative is to satisfy yourself with a *straight* chain of explanations,
starting with some principles that are taken as axioms. That's what standard "theories
of the Universe" do. What kind of "tower of explanations" you wind up with depends
on the axiomatic foundations you choose. If you are so inclined, you can take some
God (or gods) as your foundation. Or you can be introspective, realize that everything
you really know for sure about anything is what exists in your consciousness, and
take "mind" as your foundation. If you put your faith in the objectivity of physics,
you may take the laws of physics as your foundation. Even though these laws are
not in their final form, they are sturdy enough, from a practical and pragmatic point
of view, to build on them a very impressive tower. Our modern world of satellites,
computers and cell phones is a testimony to the success of this approach.

Yet, from a deep conceptual and philosophical point of view, all these foundations
suffer from a fundamental weakness: they do not seem simple and "self-evident"
enough to serve as an ultimate, "rock-bottom" foundation—some deeper level seems
required to explain them. God, at least in his more traditional incarnations, is a being
more complex than the Universe: if you take Him as your foundation, you only bury
the problem one level deeper, because now, you must explain where He comes from
and why He exists. Mind (or consciousness) also seems to be a complex, sophisticated
concept that might require some deeper level to justify its existence. As for the laws
of physics as we know them today, they are clearly not truly fundamental (they
are not even mutually compatible), although we can hope that a simpler unified
law will eventually be discovered. Even then, this law would have some *arbitrary*
characteristics, unless somehow it turns out to be the only logically possible physical
law, which is an outcome that almost no one still believes possible. Since this law
would not be a necessary, "self-evident" truth, its existence would need to be justified
by some deeper level of explanation.

As we can see, finding a fundamental "ground of being" that is truly worthy of
the name is quite a problem—we could call this the **hard problem of foundations**.
If only "nothing" could be taken as the foundation of everything[4]… what could
be simpler and more elegant, not to mention so, so Zen? Well, there might be a
way to make "nothing" into a suitable foundation, by considering something that
is equivalent to nothing: *the infinite ensemble of all abstractions.* An abstraction is

something, like a circle or the number 42, that exists without having to be embodied in a concrete way. Mathematics is the study of abstract structures, so we could speak instead of the *infinite ensemble of all mathematical structures*, or, more simply, "all-of-math". For those that have a difficult personal relationship with math that goes back to their school days, it might be strange (yet somehow comforting) to learn that all-of-math is equivalent to nothing. But it's true in a very real sense, because all-of-math contains, overall, *zero* information.[5] If you want to specify some subset of mathematics, you have to do it explicitly, and this description contains information (the bigger the subset, the more information you need to specify); but if you want to talk about the infinite ensemble of all abstractions (most of them never contemplated by any mathematician in history, of course), you can just say "all-of-math", which takes almost no time and contains essentially zero information! For me, the fact that abstractions are the most fundamental thing you can possibly imagine, and that the ensemble of all of them contains no information, makes them the ideal foundation for a theory of the Universe. I agree with science-fiction author Greg Egan when he says, "I suspect that a single 0 and a single 1 are all you need to create all universes. You just re-use them".[6]

The idea that our universe is nothing more than a mathematical structure "seen from the inside" has been called the **Mathematical Universe Hypothesis (MUH)** by Max Tegmark.[7] If the basic level of reality is an abstract mathematical structure, our universe just *has* to exist, since among all possible mathematical structures, there has to be at least one that corresponds to our world. Moreover, all mathematical structures that contain substructures that have the right properties to correspond to self-aware observers exist physically: it is the very fact that they are "perceived from within" by those self-aware substructures that makes them physical. Consequently, the MUH implies an infinite *multiverse* that contains every possible physical reality and generates every possible conscious experience: the **Maxiverse**.[8]

Several philosophers have argued that all possible worlds exist. For David Lewis, to make sense of logical statements about what could have happened in our world but did not, every possible world must be as real as ours.[9] For Robert Nozick, all possible worlds must exist on logical "egalitarian" grounds.[10] Peter Unger argues that an extreme rationalist should believe in the existence of all possible worlds, because in this case the whole of reality is less *arbitrary* than if only some worlds exist and others don't.[11]

To qualify the MUH more precisely, Tegmark often uses the name **Computational Universe Hypothesis (CUH)**. A computation is a sequential abstract structure. Since the flow of time seems to be an inescapable aspect of conscious experiences, one can make an interesting parallel between the sequential nature of computations and the apparent flow of physical time. Physicists have a fondness for whimsical acronyms, so I cannot help but propose the **Infinite Set of All Abstract Computations (ISAAC)** as a name for the basis of the CUH. To respect rigorous mathematical nomenclature, this infinite ensemble should be called a *class* instead of a *set*, but it would spoil the acronym!

Suppose that the ISAAC is the basis of all existence, and that it generates the Maxiverse. The hard problem of foundations is solved, but we now run into another one:

Fig. 11.1 The co-emergence hypothesis: abstract structures that correspond to stable, regular physical environments (φ) and those that correspond to the experiences of conscious agents (ψ) "resonate" with each other and co-emerge within the infinite set of all abstract computations ($\sum_{\infty} C$)

Fig. 11.2 A symbolic representation of Wheeler's Participatory Universe: "The universe viewed as a self-excited circuit. Starting small (thin U at *upper right*), it grows (loop of U) and in time gives rise *(upper left)* to observer-participancy—which in turn imparts 'tangible reality' to even the earliest days of the universe" (J. A. Wheeler, "Law without Law", in P. Medawar, J. H. Shelley (eds.), *Structure in Science and Art* (Elsevier, Amsterdam, 1980))

the **hard problem of lawfulness (HPL)**. If every possibility exists within the Maxiverse, irregular and chaotic worlds should greatly outnumber regular and predictable worlds like ours. Our type of universe would then be highly unlikely, which would make the Maxiverse hypothesis somewhat problematic—although David Lewis has argued that if you believe that every possible world exists, the lawfulness that we observe in our world is no more mysterious that if only one or some worlds exist.[12] Of course, one can try to solve the HPL by invoking the **anthropic principle**, the logical necessity that we observe a world regular enough to sustain our continuing existence. Somehow, this does not seem to be enough: our world is just *too* regular. Alexey and Lev Burov have argued that the observed extreme constancy of the fundamental constants of physics is hard to reconcile with the idea that our universe is a random sample within all the possible universes that could support our existence.[13]

To address the HPL, I propose to supplement the Maxiverse hypothesis with the **Co-Emergence Hypothesis** (Fig. 11.1): within the ISAAC, abstract structures that correspond to conscious agents "resonate" with each other and with abstract structures that correspond to stable, regular physical environments. This process delimitates coherent, lawful domains within the abstract space of all possibilities, the regular world that we observe being one of them. By "resonate", I have in mind something similar to Wheeler's famous analogy of the Universe as a "self-excited circuit" (Fig. 11.2). As an abstract principle that operates within the ISAAC (Fig. 11.3), co-emergence is *atemporal*: it is not a process that takes place in time, since there is no "meta-time" with respect to which the ISAAC could change or evolve.

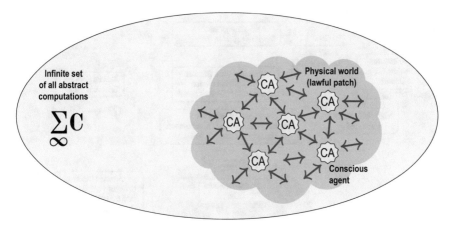

Fig. 11.3 Co-emergence operates between several individual conscious agents (CA) and their shared physical environment, delimiting a "lawful patch" within the mostly chaotic space of all possibilities

Co-emergentism has affinities with many ideas that have been proposed over the past decades as ground work towards the goal of building a physically and philosophically satisfying "theory of everything" (Fig. 11.4). The term co-emergence itself has been used by Bernard d'Espagnat to describe the relationship between states of consciousness and physical empirical reality. However, for d'Espagnat, consciousness and physical reality do not co-emerge from the set of all abstractions, but from a "veiled reality" that "lies beyond our subjective abilities at describing".[15] Co-emergence has also some similarities to what Ian Stewart and Jack Cohen call *complicity*, the process by which "two separate phase spaces join forces to 'grow' a joint phase space that feeds back into both components".[16]

Even though the ISAAC is atemporal, in the physical universes that exist within it, conscious observers perceive the flow of time: the concepts of causation and causality can be applied. For co-emergence to make sense, it is beneficial to extend the notion of causality to include both directions of time, hence the relevance of the ideas of Huw Price about retrocausality.[17] As noted by Ken Wharton,[18] the Lagrangian formulation of physics, in terms of path integrals and stationary action, can offer valuable insights about the deep logical structure of our world: in a "Lagrangian Schema Universe", explanations need not always be in the Newtonian form "from t to $t + dt$", which leaves room for two-way causality and co-emergence.

Emergence is usually understood in terms of properties of a system that exist at a higher level of description and have no equivalent at a lower level: one classic example is the *fluidity* of water, which has no meaning at the level of the individual molecules. In the co-emergence of a physical lawful environment and the community of conscious agents that observes it, emergence works both ways. Consciousness, with its power of agency and volition, emerges out of a physical level of description where interactions take place according to "mindless" laws, while the rigid laws that

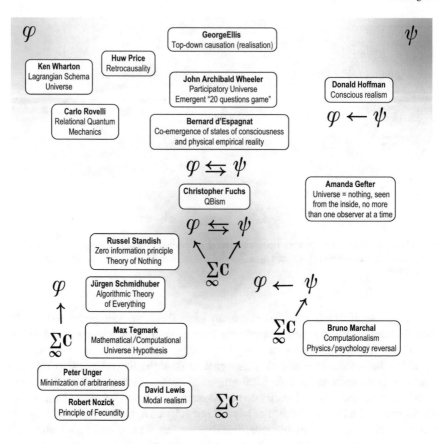

Fig. 11.4 The co-emergence of co-emergentism

obey the physical interactions are, in some real sense, an emerging consequence of
the existence of a community of conscious observers that share between themselves
a coherent story about a lawful and stable world. Current and future research on the
topic of top-down causation (or "realisation"), by George Ellis[19] and others, can help
in understanding the details of how co-emergence operates within the ISAAC.

My conception of co-emergence has been greatly influenced by the ideas of
Russell Standish, himself elaborating on the work of Bruno Marchal[20] and Jürgen
Schmidhuber.[21] In his book "Theory of Nothing",[22] Standish writes:

> Consciousness (…) exists entirely in the first-person perspective, yet by the Anthropic Prin-
> ciple, it supervenes on (or emerges out of) first person plural phenomena. (…) However, we
> also have the third-person world emerging out of consciousness (…) The Anthropic Principle
> cuts both ways—reality must be compatible with the conscious observer, and the conscious
> observer must supervene on reality.

The tension between an objective, third-person description of the world, and a
subjective, first-person description, is of course at the heart of the difficulties physi-

cists have been having, for almost a century now, to give a satisfying interpretation to quantum mechanics—especially to the "projection postulate" that links the quantum world, evolving unitarily according to the Schrödinger equation, and the classical world where we always observe a single outcome for a particular experiment. Of all the interpretations of quantum mechanics, QBism, a relative newcomer, "resonates" particularly well with co-emergentism. In QBism, every observer has his own wave-function, which is a description of his own knowledge or belief about the system. According to QBism, quantum mechanics is a theory of the relationship between each observer and the physical world. In the words of Christopher Fuchs, one of the main proponents of QBism,

> Quantum mechanics is a single-user theory, but by dissecting it, you can learn something about the world that all of us are immersed in. (…) it's not that the world is built up from stuff on "the outside" as the Greeks would have it. Nor is it built up from stuff on "the inside" as the idealists (…) would have it. Rather, the stuff of the world is in the character of what each of us encounters every living moment—stuff that is neither inside nor outside, but prior to the very notion of a cut between the two at all.[23]

According to Fuchs, QBism, as well as related interpretations of quantum mechanics like Relational Quantum Mechanics, developed by Carlo Rovelli,[24] should be labeled *participatory realism*, an homage to the "participatory universe" idea of John Archibald Wheeler.

In Table 11.1, I consider six more-or-less "hard" problems of physics/metaphysics, and I contrast how co-emergentism addresses them with the way they can be addressed by general theories of the Universe based on other foundations. Unfortunately, within the scope of this article, there is not enough space to discuss in detail all the entries in the table. I have already mentioned the hard problems of foundations, lawfulness, and of the interpretation of the projection postulate in quantum mechanics. The problem of *free will and effective intention* is an interesting one. In every day life, we experience the first-person perspective of being a conscious *agent*: we have goals, act with intention and have an impression of free will. We believe that our goals, intentions and willful actions have an effective causal impact on what happens in the world. Of course, in most theories of the Universe, if we consider the whole of reality and we do not allow for a "meta-time" with respect to which this "whole" can change or evolve, our goals, intentions and free will cannot be *globally* meaningful, even if they *locally* mean something to us. However, in co-emergentism, the properties of the local lawful physical patch that conscious agents find themselves in is co-determined by the actions of the agents, so one can argue that goals, intentions and free will, even if they are still globally meaningless, somehow acquire more significance.

The *problem of delusion* would deserve an article of its own. The possibility that "deluded" observers (simulated beings within the computers of advanced civilizations, or freakish "Boltzmann brains" fluctuations) outnumber "non-deluded" ones has recently kept some physicists awake at night. In the Maxiverse spanned by the ISAAC, there are of course an infinite number of deluded observers, and an infinite number of non-deluded ones. But if co-emergentism is true, what really matters is the immediate, local relationship between the community of conscious observers

Table 11.1 Hard problems and where to find them

Problem of	God first	Mind first	Physics first	Math first	Co-emergentism
Foundations What is the fundamental level of existence?	Moves the problem one level deeper: what explains God's existence?	"Mind" might be too complex to be the fundamental level	Why these laws? Why these initial conditions?	Easy! Math is abstract, abstractions simply are	The infinite set of all abstract computations
Lawfulness Can we explain why our world obeys laws with such implacable regularity?	God: "I am the law!"	If you start with sane mind(s), you get a lawful world… but what about insane minds?	Easy… if you take for granted the laws of physics	Hard problem! If every possible world and conscious experience exists within "all-of-math", shouldn't most be chaotic?	Lawfulness is a local "resonance" defined by the interplay of conscious agents and their physical environment
Free will and effective intention Do our goals, intentions and impression of free will have an effective causal effect on what happens?	God allows it if He so pleases	For all we know, free will and the ability to act intentionally towards goals might be a basic attribute of consciousness	In any real sense, probably not (and quantum randomness does not help), but you can console yourself with compatibilism	No, because no matter what, everything happens to some version of you in the Maxiverse anyway	Might be locally significant within our co-emergent "lawful patch", even if globally, everything still happens anyway
Interpretation of the projection postulate How does the quantum wave-function "transition" to the observed classical world?	Maybe God made quantum mechanics to annoy physicists, or keep them occupied forever	The problematic "intrusion" of the observer in quantum mechanics makes "mind first" more believable	It's been almost a century and we still don't know!	Same problem as "physics first"	The problematic "intrusion" of the observer in quantum mechanics might be a sign that co-emergentism is on the right track

(continued)

Table 11.1 (continued)

Problem of	God first	Mind first	Physics first	Math first	Co-emergentism
Delusion Can we know that the world we observe in our waking lives is not a charade or a prank?	A "fair-play" God would not allow his creatures to be deceived (Descartes' argument)	A conscious experience can never be *wrong* in itself	Simulation and Boltzmann brain problems	Same problem as "physics first", exacerbated by the intractable measure problem in the infinite Maxiverse	If you try to push your reasoning too far away from your observed reality, it may no longer apply
Solipsism Can I be reasonably sure that I share a world with other conscious beings?	If God does not deceive us, there are other conscious beings in the world	Always a possibility… maybe we are all one mind anyway	If we are not deluded, other humans, being physically identical to me, should equally be conscious	In all-of-math, there are isolated structures that are effectively "solipsistic minds", but their proportion is hard to evaluate (measure problem)	The lawfulness of the physical environment co-emerges via the relationship between conscious agents, even if each conscious agent has his own irreducible viewpoint

and the physical reality they observe. In Fig. 11.3, if you move too far out, the cloud that symbolizes our lawful patch dissolves into the relative chaos that characterises most of the ISAAC: physics becomes indeterminate, or most likely simply irrelevant. Could it be that, when we worry about the proliferation of deluded observers, we try to push our reasoning too far away from our observed reality, into a realm where it no longer applies? In the same way, could the dead-ends we have been encountering over the past decades in fundamental physics (the failure to unify quantum mechanics and general relativity, the proliferation of solutions in the landscape of M-theory) be interpreted as signs that we are nearing the edge of our patch of lawfulness in the space of all possibilities?

The *problem of solipsism* would also deserve an article of its own—in a form given "new life" by the recent developments in physics, like QBism and the black hole firewall paradox.[25] No sane physicist actually argues that he is the only conscious being in the universe. It's just that some fundamental coherence problems arise when we try to combine the first-person viewpoints of different observers into a single third-person "truly objective" reality. It is as if physics is trying to tell us that the world

Fig. 11.5 You drive alone, at night, on a desert road. The sky is full of stars. Suddenly, you see a sign by the side of the road…

arises out of the point of view of single observers, even if they do in the end form a community that observes a single unified reality (Fig. 11.5). Of course, elucidating the relationship between first-person singular, first person plural and third-person point of views is of crucial importance if we hope to clearly articulate the meaning of co-emergentism.

For now, co-emergentism is only a working hypothesis. Like many speculative hypotheses concerning the foundational questions of existence, it hasn't yet reached the point where it can claim to have strong results or to make detailed predictions. In other words, it does not have a "shut-up and calculate" aspect that can provide reasonably comfortable day jobs for physicists. But research continues, and things might change. Donald Hoffman has been exploring a working hypothesis he calls *conscious realism*: he takes consciousness as the ground of existence, and is trying to make physics emerge out of the interaction between conscious agents, by applying a generalized abstract form of the principles of natural selection.[26] It is an ambitious enterprise, but if it succeeds, it could provide a starting point for developing a fully-fledged theory of how conscious agents can co-emerge alongside their environment, and make their little corner of the Maxiverse a safe, cozy place to call home.

End notes

1. C. A. Fuchs, "On participatory Realism", in I. T. Durham, D. Rickles (eds.), *Information and Interaction: Eddington, Wheeler and the Limits of Knowledge* (Springer 2017)
2. J. A. Wheeler, "Genesis and Observership," in *Foundational Problems in the Special Sciences: Part Two of the Proceedings of the Fifth International Congress of Logic, Methodology and Philosophy of Science, London, Canada* (1975)
3. J. A. Wheeler, "Frontiers of Time," in *Problems in the Foundations of Physics, Proceedings of the International School of Physics* (North-Holland, Amsterdam, 1979)
4. A. Gefter, *Trespassing on Einstein's Lawn* (Bantam Books, New York, 2014)
5. R. K. Standish, *Theory of Nothing* (BookSurge, Charleston, S.C, 2006)
6. J. G. Byrne, "Interview: Greg Egan, Counting Backwards from Infinity", Eidolon magazine, issue 15 (July 2004)

7. M. Tegmark, *Our Mathematical Universe* (Alfred A. Knopf, New York, 2014)

8. M. Séguin, "My God, It's Full of Clones", in A. Aguirre, B. Foster, Z. Merali (eds.), *Trick or Truth? The Mysterious Connection Between Physics and Mathematics* (Springer, 2016)

9. D. Lewis, *On the Plurality of Worlds* (Blackwell, Oxford, 1986)

10. R. Nozick, *Philosophical Explanations* (Harvard University Press, 1981)

11. P. Unger, "Minimizing Arbitrariness", Midwest Studies in Philosophy, IX (1984)

12. see note 9

13. A. Burov and L. Burov, "Genesis of a Pythagorean Universe", in A. Aguirre, B. Foster, Z. Merali (eds.), *Trick or Truth? The Mysterious Connection Between Physics and Mathematics* (Springer, 2016)

14. J. A. Wheeler, "Law without Law", in P. Medawar, J. H. Shelley (eds.), *Structure in Science and Art* (Elsevier, Amsterdam, 1980)

15. B. d'Espagnat, *On Physics and Philosophy* (Princeton University Press, 2006)

16. I. Stewart and J. Cohen, *Figments of Reality* (Cambridge University Press, 1997)

17. H. Price, "Does Time-Symmetry Imply Retrocausality?", in Studies in History and Philosophy of Modern Physics 43 (2012)

18. K. Wharton, "The Universe is not a Computer", in A. Aguirre, B. Foster, Z. Merali (eds.), *Questioning the Foundations of Physics* (Springer, 2015)

19. G. Ellis, "Recognising Top-Down Causation", in A. Aguirre, B. Foster, Z. Merali (eds.), *Questioning the Foundations of Physics* (Springer, 2015)

20. B. Marchal, *Calculabilité, Physique et Cognition*, PhD thesis, Université de Lille (1998) http://iridia.ulb.ac.be/~marchal/

21. J. Schmidhuber, "Algorithmic theories of everything", *Technical Report IDSIA-20-00, IDSIA*, Galleria 2, 6928 Manno (Lugano), Switzerland (2000) arXiv:quant-ph/0011122

22. see note 5

23. A. Gefter, "A Private View of Quantum Reality" (interview with Christopher Fuchs), Quanta Magazine (2015)

24. C. Rovelli, "Relational Quantum Mechanics", International Journal of Theoretical Physics 35 (1996)

25. see note 4

26. D. Hoffman and C. Prakash, *Objects of consciousness*, Frontiers in Psychology (2014)

Acknowledgements The author would like to thank Susan Plante and Julie Descheneau for reviewing early drafts of this manuscript, and for many fruitful discussions.

Chapter 12
God's Dice and Einstein's Solids

Ian T. Durham

What does chance ever do for us?

—William Paley (1743–1805)

What role does chance play in the universe? Quantum theory suggests that randomness is a fundamental part of how the universe works and yet we live mostly intentional, ordered lives. We make decisions with the expectation that our decisions matter. How is it possible for this directed and seemingly deterministic world to arise from mere randomness? In this essay I show that simple combinatorial systems behave in precisely this manner and I discuss the implications of this for theories that include notions of free will.

12.1 Process and Chance

What role does chance play in the evolution of the universe? Until the development of quantum mechanics, the general consensus was that what we perceived as chance was really just a manifestation of our lack of complete knowledge of a situation. In a letter to Max Born in December of 1926, Einstein wrote of the new quantum mechanics, "The theory says a lot, but does not really bring us any closer to the secret of the 'old one.' I, at any rate, am convinced that *He* does not throw dice." [6]. Yet all attempts to develop a deterministic alternative to quantum mechanics have thus far failed. At the most fundamental level, the universe appears to be decidedly random. How is it, then, that the ordered and intentional world of our daily lives arises from this randomness? Intuitively, we tend to think of randomness as being synonymous with unpredictability. Of course the very nature of the term 'unpredictable' implies agency since it implies that an agent is actively making a prediction. I am not interested here

I. T. Durham (✉)
Department of Physics, Saint Anselm College, Manchester, NH 03102, USA
e-mail: idurham@anselm.edu

© Springer International Publishing AG, part of Springer Nature 2018
A. Aguirre et al. (eds.), *Wandering Towards a Goal*, The Frontiers Collection,
https://doi.org/10.1007/978-3-319-75726-1_12

in whether or not the universe requires an agent to make sense of it. I'm interested in understanding if it is possible for intentionality to arise naturally from something more fundamental and less ordered regardless of whether that process requires an agent.

While we typically think of the universe as being a collection of 'things'—particles, fields, baseballs, elephants—it is the processes that these things participate in that make the universe interesting. A process need not require an active agent. A process is simply a change in the state of something where by 'state' we simply mean a set of properties and (possibly) correlations about the system. This definition of process is similar to the concept of a *test* from operational probabilistic theories (see [2] for a discussion of such theories). Similar to such theories, then, we can define a **deterministic** process as being one for which the outcome can be predicted with certainty. To put it another way, a deterministic process is one for which there is only a single possible outcome. A random process must then be an *unpredictable* change. That is, a process for which there is more than one possible outcome is said to be **random** if all possible outcomes are equally likely to occur. It's important to note that there is a difference between the *likelihood* of making an accurate prediction of the outcome of a process and one's *confidence* in that prediction. Confidence can be quantified as a number between 0 (no confidence) and 1 (perfect confidence). Likelihood is just the probability that a given outcome will occur for a given process. The outcomes of random processes are all equally likely to occur and, in such cases, one's confidence in accurately predicting the correct outcome should be zero.

The definition of determinism given here differentiates it from the concept of causality. D'Ariano, Manessi, and Perinotti have argued that confusing the two ideas—determinism and causality—has led to misinterpretations of the nature of EPR correlations [4]. Determinism and randomness represent the extremes of predictability. A fully causal theory can have both. Just because a process is random does not mean the outcome of the process doesn't have a cause. Conversely, as D'Ariano, Manessi, and Perinotti have shown, it is possible to have deterministic processes *without* causality. The point is that determinism is associated with the likelihood of outcomes for a given process whereas causality is associated with the fact that each process has an input state and an output state.

Of course, many real processes are neither random nor deterministic. There may be more than one possible outcome for a process, but those outcomes may not be equally likely to occur. What do we call such processes? For a two-outcome process whose outcomes have a 51% and 49% likelihood of occurrence respectively, one might be tempted to refer to it as 'nearly random.' On the other hand, if those same likelihoods were 99% and 1% respectively, one might be tempted to say the process was 'nearly deterministic.' But what if they were 80 and 20% or 60 and 40%? At what point do we stop referring to a process as 'nearly deterministic' or 'nearly random'? We need less arbitrary language. One suggestion would be to refer to such in-between cases as 'probabilistic.' But this is misleading since we can still assign probabilities to the outcomes of both random and deterministic processes; they are no less probabilistic than any other process.

A solution presents itself if we consider the aggregate, long-term behavior of such processes. As an example, consider that casinos set the odds on games of craps—a game that is neither random nor deterministic—under the assumption that they will make money on these games in the long run and (crucially) that the amount of money they will make is reliably predictable within some acceptable margin of error. So the process of rolling a pair of dice (which is all that craps is) is at least partially deterministic to a casino. But now consider a game with two outcomes, A and B, whose respective probabilities of occurring are 50.5 and 49.5%. Could a casino set up a system by which they could, within some margin of error, make a long-term profit on this game, even if that profit is very small? Suppose it costs $100 to play this game and that a player receives $102 if outcome B occurs but nothing if outcome A occurs. Suppose also that, on average, the casino expects 10,000 people to play this game each year. That means that, on average, they will pay out $504,900 a year in winnings but keep $505,000 a year in fees leaving them with $100 in profit (on average). Though this is ridiculously low, the crucial point is that *it is not zero*. As low as it is, the casinos can still budget for it and, in the long run, can expect to make a profit on it. The fact is that something like this can be done for *any* process that is not fully random and yet not fully deterministic, but the proof is physical rather than mathematical.

This is an important point that is sometimes misunderstood. Suppose that we have a process with two outcomes, $\mathcal{O}_1$ and $\mathcal{O}_2$, whose probabilities are p_1 and p_2 respectively. This process could be fully random, fully deterministic, or something in between. According to the law of large numbers, if this process occurs n times, as $n \to \infty$ we will find that the probability of occurrence for $\mathcal{O}_1$ is p_1 while for $\mathcal{O}_2$ it is p_2. Crucially, this does *not* mean that we are *guaranteed* to find exactly np_1 occurrences of $\mathcal{O}_1$. The law of large numbers—and, indeed, any probabilistic law or theory—merely makes exact predictions about *probabilities*. Whether or not the universe obeys those laws and to what degree, is a physical matter. As it so happens, no statistically significant *physical* violation of the law of large numbers is known to exist. But this applies equally well to both fully random as well as fully deterministic processes. Just because a particular outcome for a given process has a probability equal to unity (a mathematical statement) does not mean it is guaranteed to occur (a physical result). Likewise just because all possible outcomes of a process are equally likely (a mathematical statement) does not mean that, in the aggregate, a large number of these processes will evenly distribute the outcomes (a physical result). In short, the physical universe is not *required* to obey any particular mathematical laws. The fact that it *does* is technically unexplained by either mathematics or physics, as of now.

In lieu of this, I propose to call the processes in between fully random and fully deterministic, **partially deterministic** since we fully expect, as casinos do, that physical systems obey the law of large numbers (though our only proof is that no meaningful violation has ever been observed) and thus that such in-between processes are predictable *in the aggregate*.

12.2   God's Dice

The aforementioned game of craps simply involves betting on the outcome of a roll of a pair of dice. The game is as old as dice themselves and serves as a useful example of how some level of partial determinism can arise from randomness. It also provides a straightforward method for introducing a few additional terms. Those wishing to delve more deeply into this subject are encouraged to dive into Refs. [1, 9, 10].

Consider a fair, six-sided die. As a fair die, it is assumed that upon rolling this die, all of the six outcomes are equally likely. In fact casinos paint the dots on their dice, rather than use the usual divots because the divots are not equally distributed and thus throw off the center-of-mass which changes the long-term probabilities.[1] While real dice are never truly random, a so-called 'fair die' is considered to be a theoretical ideal and is thus random.

Now consider a roll of two fair dice as in a game of craps. Since they are both fair dice, each outcome on each individual die is equally likely. We are also assuming that we can easily distinguish the dice from one another e.g. perhaps one is blue and one is red. Considered together, then, there are thirty-six possible outcomes—configurations—to a single, simultaneous roll of both. Since we can distinguish between the two dice, if the roll produces a four on the blue die and a three on the red die, this is an entirely different outcome from a three on the blue die and a four on the red one. Each of these configurations is referred to as a *microstate*.

But in craps, as in other games that use a pair of dice, we are often interested in the *sum* of the numbers on the faces. Thus we typically consider the roll of a pair of dice as giving us a number between two and twelve. We call this number the *macrostate*. If we look at each of the thirty-six microstates, we'll see that they can be grouped according to which macrostate they produce. The number of microstates that will produce a given macrostate is known as the *multiplicity* and is given the symbol Ω. But note that the multiplicities of the macrostates for the roll of a pair of fair dice are not all equal. There are, for instance, six different combinations that can produce a roll of seven. On the other hand, there is one and only one way to roll a two or a twelve.

The probability of a given roll (i.e. macrostate) is given by the multiplicity of that roll divided by the *total* multiplicity. So, for example, the probability of rolling a seven is six divided by thirty-six or one-sixth. Conversely, the probability of rolling a two or a twelve is one-thirty-sixth. Table 12.1 lists the microstates for each macrostate of the pair of dice, giving the multiplicity and probability of each. Though we think of this as a single roll of a pair of dice, it is really two simultaneous rolls of individual dice. Each of these individual rolls is a random process in that the probability of an individual die yielding any one of its six possible outcomes is precisely the same. Yet when the pair are considered simultaneously, the combined outcome is partially deterministic. As should be clear from Table 12.1 this behavior is not physical in the sense that the probabilities of the individual macrostates are due to the combinatorics of the problem or what one might call 'mindless' mathematics. So a pithy counter

---

[1] They also routinely replace their dice since the sides of dice can wear unevenly. See Ref. [7].

Table 12.1 This table lists the microstates for each macrostate for a roll of a pair of six-sided dice. The multiplicity, Ω, is the total number of microstates

Macrostate	Microstates	Ω	Probability
2	⚀⚀	1	1/36
3	⚀⚁, ⚁⚀	2	$2/36 = 1/18$
4	⚀⚂, ⚁⚁, ⚂⚀	3	$3/36 = 1/12$
5	⚀⚃, ⚁⚂, ⚂⚁, ⚃⚀	4	$4/36 = 1/9$
6	⚀⚄, ⚁⚃, ⚂⚂, ⚃⚁, ⚄⚀	5	5/36
7	⚀⚅, ⚁⚄, ⚂⚃, ⚃⚂, ⚄⚁, ⚅⚀	6	$6/36 = 1/6$
8	⚁⚅, ⚂⚄, ⚃⚃, ⚄⚂, ⚅⚁	5	5/36
9	⚂⚅, ⚃⚄, ⚄⚃, ⚅⚂	4	$4/36 = 1/9$
10	⚃⚅, ⚄⚄, ⚅⚃	3	$3/36 = 1/12$
11	⚄⚅, ⚅⚄	2	$2/36 = 1/18$
12	⚅⚅	1	1/36
	Total:	36	1

to Einstein's objection might be that a dice-throwing God still produces partially predictable results.

There is one objection to this example that is worth considering. The numbers on the dice are entirely arbitrary. That is, we could have instead painted six different animals on the faces of each die. In this case we might find it hard-pressed to identify any distinctive macrostates other than pairs and we could eliminate the pairs by painting different animals on each die. Thus it seems as if the macrostates used in typical die rolls are entirely arbitrary in the sense that their relative import is based on a meaning that we *assign* to them. The labelling of the sides of the dice is not a fundamental property of the dice themselves. We can get around this problem and improve on our odds by perhaps ironically considering a model proposed by Einstein nineteen years before his comment to Born.

12.3   Einstein's Solids

In 1907 Einstein proposed a model of solids as sets of quantum oscillators. That is, each atom in such a solid is modeled in such a way that it is allowed to oscillate in any one of three independent directions. Thus a solid having N oscillators would consist of $N/3$ atoms. Crucially, since the oscillators are quantum, they can only hold discrete amounts of energy. So, for instance, suppose that we have an overly simplified Einstein solid containing just a single atom and thus three oscillators. If we supply that solid with a single discrete unit of energy, that energy unit could be absorbed by any one of the three oscillators, but cannot be further subdivided and shared among them. It is assumed that the process of absorption is random. In other words, for a macrostate consisting of a single unit of energy, there are three equally likely microstates, i.e. $\Omega = 3$. In general, for an Einstein solid with N oscillators and q units of energy, the multiplicity is

$$\Omega(N, q) = \binom{q + N - 1}{q} = \frac{(q + N - 1)!}{q!(N - 1)!}. \tag{12.1}$$

Now consider two Einstein solids that are weakly thermally coupled and approximately isolated from the rest of the universe. By weakly thermally coupled, I mean that the exchange of thermal energy between them is much slower than the exchange of thermal energy among the atoms within each solid. This means that over sufficiently short time scales the energies of the individual solids remain essentially fixed. Thus we can refer to the macrostate of the isolated two-solid system as being specified by the individual fixed values of internal energy. (For a further discussion, see Ref. [10].) Let's begin by considering a simple (albeit unrealistic) system. Suppose each of our two solids has three oscillators, i.e. $N_A = N_B = 3$, and the system has a total of six units of energy that can be divvied up between the oscillators. Suppose that we put all six of these units of energy into solid B. That means that there is only one possible configuration for the oscillators in solid A—they all contain zero energy. Conversely, there are twenty-eight configurations for the oscillators in solid B according to (12.1).[2] The total number of configurations for the system as a whole is just the product of the two and thus is also twenty-eight.

Suppose that we instead put a single unit of energy into solid A with the rest going to solid B. In this case, there are three possible configurations for solid A since the single unit of energy we've supplied to it could be in any one of the three oscillators. The five remaining units of energy can be distributed in any one of twenty-one ways within solid B. But now the *total* number of configurations for the system is $3 \cdot 21 = 63$. Table 12.2 summarizes the energy distribution and corresponding multiplicity for this simple system and Fig. 12.1 shows a smoothed plot of the total multiplicity, Ω_{tot} as a function of the energy q_A contained in solid A. This tells us that the states for which the energy is more evenly balanced between the two solids are more likely to occur because there are more possible ways to distribute the energy in such cases. This is analogous to the example given in the previous section involving a pair of fair dice. There is no intentionality on the part of the system. In addition, the system is considered to be isolated from the rest of the universe and thus there is no environment driving these results. They are simply due to combinatorics. This does not *guarantee* that our results will follow these predictions precisely, but as I mentioned previously, there is no known violation of such mathematical laws by

---

[2]Note that (12.1) holds for solid A as well since $0! := 1$.

Table 12.2 This table shows the distribution of six units of energy among two Einstein solids, each with three oscillators

q_A	Ω_A	q_B	Ω_B	$\Omega_{tot} = \Omega_A \Omega_B$
0	1	6	28	28
1	3	5	21	63
2	6	4	15	90
3	10	3	10	100
4	15	2	6	90
5	21	1	3	63
6	28	0	1	28

Fig. 12.1 This shows a smoothed plot of Ω_{tot} as a function of q_A for the data from Table I

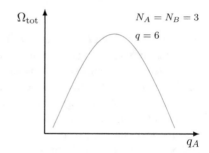

any physical system. Each individual microstate of the combined system is assumed to be equally probable and thus the process of reaching one of these microstates from any other is completely random. It just happens that more of those microstates correspond to configurations in which the energy is more evenly divided between the two solids. Thus the system can undergo random fluctuations about the mean and still be more likely to be found in a microstate in which the energy is roughly equally divided between the two solids.

But consider now what happens when we begin to scale the system up to more realistic sizes. Figure 12.2a shows a plot of the total multiplicity of the system as a function of the energy in solid A when the total number of oscillators and the total number of energy units is a few hundred. Figure 12.2b shows a plot of the same function when the number of oscillators and energy units is a few thousand. The larger our Einstein solids become, the narrower the peak of the multiplicity function. For realistic Einstein solids, the peak is so narrow that only a tiny fraction of microstates have a reasonable probability of occurring. That is, *random fluctuations away from equilibrium are entirely unmeasurable.*

It is important to keep in mind that all we have done in Fig. 12.2 is to scale up the Einstein solids. Each individual microstate remains equally probable and thus the underlying process of moving from one microstate to another is entirely random. Yet, as the system grows larger and larger, it is increasingly likely to be found in only one of a very small number of macrostates. This means that we can make *highly* accurate

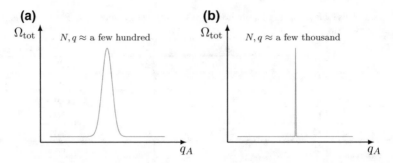

Fig. 12.2 a This shows a plot of Ω_{tot} as a function of q_A for a pair of Einstein solids when $N, q \approx$ a few hundred. **b** This is the resultant plot for the solids when $N, q \approx$ a few thousand

predictions of which macrostates will occur given some initial input data. This is a dramatically scaled up analogy to a game of craps. Though the fluctuations taking the system from one microstate to another are entirely random, the *macrostate* is very nearly deterministic. Thus we have a situation in which a nearly deterministic process can arise from a collection of random processes. Due to the fact that no statistically significant violation of the law of large numbers has ever been observed, we are justified in assuming that this mathematical prediction should hold for all physical systems. In addition, unlike the situation with the dice, we are not arbitrarily assigning meaning to the microstates and macrostates.

An obvious question is whether or not it is possible to achieve *perfect* determinism with a pair of Einstein solids. Certainly in the limit that $N, q \rightarrow \infty$ the narrowness of the multiplicity peak in our example becomes asymptotically thin. The limit in which a system becomes large enough that random fluctuations away from equilibrium become unmeasurable is known as the *thermodynamic limit*. In other words, at some point our partially deterministic system becomes indistinguishable from a fully deterministic one. Where that transition occurs may depend on a host of factors, but the reason it is referred to as the thermodynamic limit is precisely because it is where conventional thermodynamic methods of analysis—which are deterministic!—become the most useful way to understand the behavior of a pair of solids that are in thermal contact with one another.

Of course, this is just a single example from one area of physics but it serves to show that near-perfectly deterministic macroscopic processes *can* arise from a very large number of random microscopic processes.

12.4 Process and Free Will

Let us consider a toy universe in which all microscopic processes are random and thus equally probable. The only physical constraints that we will place on this toy universe are to (1) limit the outcomes of each microscopic process to being finite in number, (2) require that these outcomes be distinguishable from one another, and (3) require that the physical outcomes match the mathematical predictions to a high degree of accuracy, i.e. there will be no statistically significant physical deviations from mathematical predictions, at least in the realm of combinatorics and probability. Macroscopic processes in such a toy universe would have varying levels of determinism based on the combinatorics and the nature of the processes themselves. For example, the microstates and macrostates of a pair of six-sided dice are different from the microstates and macrostates of the combination of one six-sided die with one eight-sided die. Thus the nature of the dice dictate which processes are allowed in each case (e.g. a roll of fourteen is not possible with a pair of six-sided dice). For our toy universe, we can think of any constraints as being dictated by the initial (physical) conditions of the universe itself.

It is worth asking, then, what it would mean for a hypothetical 'being' in such a universe to have free will. Free will is generally viewed as one's ability to freely choose between different courses of action.[3] This requires, however, that when presented with a choice, an agent can reliably predict the outcome of some process. If I am, for instance, faced with the choice of carrots or broccoli as a vegetable side for my dinner, the essence of free will is that, free of unpredictable external factors, if I choose to have carrots I can have confidence that I will actually have carrots with my dinner, i.e. the carrots won't randomly and inexplicably turn into a potato the moment they touch my plate. The crucial but subtle difference here is that my choice in this example is between two *different processes*—the process of physically taking carrots from my refrigerator or the process of physically taking broccoli from my refrigerator—rather than two different outcomes of a *single* process. So once I have chosen to carry out one or the other of these processes, I can have confidence that the multiplicity of one outcome of my chosen process is so much greater than the multiplicity of any other outcome that my desired result will actually occur, i.e. the probability of the most likely macrostate *not* occurring is utterly unmeasurable.

Of course, any beings in our hypothetical toy universe are unequivocally *part* of that universe and thus an amalgam of random processes themselves. If the deterministic macroscopic processes arise from microscopic random ones solely due to the combinatorics of a large number of such microscopic processes, then it is worth asking if free will really does exist. This is certainly a fair question, but misses the broader point. Regardless of what happens at the most fundamental level, the concept of free will is meant to be applied to sentient beings (which are inherently *not* fundamental) making conscious choices about the macrostates of large-scale systems. As sentient beings we expect that free will entails our ability to freely make a choice with the confidence that a specific outcome of our chosen process really does occur with a high degree of probability. For that to happen certain processes must be at least partially deterministic if not fully so.

This brings up an important distinction. There are really different *levels* of processes. We can refer to a process associated with a macrostate as a *macroprocess*.

---

[3]I recognize that there are many, often divergent notions of free will. I am focusing on theories that include a generally colloquial notion of the concept.

The constituent processes of a macroprocess would then be *microprocesses*. The macroprocess of simultaneously rolling a single pair of dice is composed of two microprocesses—the independent rolls of two individual dice.[4] So the terminology refers to the level of the system and not necessarily the size of the system or its constituents. The act of me pulling carrots out of a bin in my refrigerator is a macroprocess that actually consists of trillions of microprocesses involving the neurons in my brain, the electrical signals in my neurological system, the mechanical motion of the refrigerator parts, etc. These in turn are all made up of further constituent processes all the way down to the processes involving the fundamental particles and fields that constitute the material foundation of the entire system.

Free will thus generally involves choices about *macroprocesses* with varying degrees of confidence.[5] I may be highly confident that the carrots in my refrigerator won't spontaneously turn into potatoes, but I'm a tad less confident that inserting the key into the ignition of my car will turn the car on. Certainly I expect it to turn on *most* of the time, but it is entirely plausible that something could go wrong and it won't turn on. I'm even less confident when I approach an unfamiliar intersection and don't know which way to go. Depending on the situation, my choice could essentially be entirely random. The key point here is that if *all* macroprocesses were entirely random, we wouldn't even have the *illusion* of free will because our choices would be meaningless since they would be based entirely on guesses. So any concept of free will requires that most *macroprocesses* be at least partially deterministic. But, crucially, *microprocesses* can still be random since their combinatorial behavior can lead to partially deterministic macroprocesses like the rolling of dice or the equilibrium state of two solids in thermal contact.

12.5   Boundary Conditions

There's one final objection to this line of argument that should be addressed. It's clear that the emergence of determinism and free will in this model is not solely due to the combinatorics alone. As I have said, we are assuming that the physical systems of interest will obey the mathematical rules we are employing. We make this assumption in good faith given the fact that no violations have ever been observed. Likewise, at the most fundamental level there has to be something non-mathematical in order to distinguish, for example, a quark from a lepton or even the number one from the number two. But it is worth asking if the combinatorics itself can produce additional boundary conditions on the system that then further constrain its evolution. In other

---

[4]It is worth mentioning here that if the rolls are *not* independent (i.e. they are in some way correlated), this merely alters the number of possible outcomes of any given process. It doesn't change the definitions or the conclusion of this analysis in any way.

[5]This is even true in quantum settings. Free will in a quantum setting is typically interpreted as having to do with a choice of measurement basis which is usually tied to a physically *macroscopic* measuring device. It has nothing whatsoever to do with the microscopic quantum processes that produce the outcome of those measurements.

words, is it possible for a system's own internal combinatorics, i.e. pure mathematics (assuming the physical system hews to the mathematics) to change the probabilities of future macrostates?

In the simple example using dice, no matter how many times we roll them, the combinatorics alone will not change the probabilities of the macrostates. Certainly the dice could wear down unevenly over time, but this is a physical effect. But consider a pair of Einstein solids in thermal contact as I described in the previous section. A microprocess for such a system is the shifting of an energy unit from one oscillator to another. This microprocess is fundamentally random. If we introduce a large number of energy units to such a system and assume it has a large number of oscillators, regardless of how those energy units are initially distributed, over time the system will find itself limited to just a few possible macrostates. Crucially, these random microprocesses don't suddenly cease to occur when the system reaches equilibrium. Energy continues to be passed around while the underlying microprocesses remain random, yet fluctuations away from equilibrium eventually become unmeasurable. This is simply because a few macrostates near equilibrium have a much higher probability of occurring than all the other macrostates. In this sense, the macrostate corresponding to equilibrium has imposed a boundary condition on *future macroprocesses* purely through combinatorics (and the assumption that the physical system will obey the combinatorial prediction). So while the evolution toward equilibrium has no effect on a system's underlying microprocesses, which are presumably fixed by the inescapable laws of physics, it *can* have an effect on the future evolution of the *macroprocesses* due to the system's own internal combinatorics.

12.6 Conclusion

There is little in this essay that is actually speculative. Admittedly I am considering highly simplified systems here, but they at least demonstrate that it is possible for something ordered and intentional to arise from the aggregate behavior of a collection of random processes with no external forcing, i.e. due solely to combinatorics and assuming that physical systems obey the mathematical rules that are applied. The behavior of such systems also suggests that it is entirely possible for free will, at least as it is colloquially understood, to emerge from something far less ordered. In fact both Eddington and Compton argued that the randomness of quantum mechanics was a *necessary condition* for free will [3, 5]. On the other hand, Lloyd has argued that even deterministic systems can't predict the results of their decision-making process ahead of time [8]. Is free will just an illusion? Does it require randomness or does it require determinism? The answers to these questions undoubtedly lie in a deeper understanding of the transition from quantum systems to classical ones. In this essay I have shown that the seeds of such an understanding might be found in simple combinatorics. The mindless laws of mathematics might be just what allows the universe to evolve intentionality. At the very least, it is worth a deeper look.

Acknowledgements I would like to thank Irene Antonenko for pressing me on the language of partial determinism. I stand by my use of the term, but Irene's comments helped me to clarify why I prefer it. Plus it made for a great after-dinner discussion that was enhanced by good dessert and good wine. I additionally thank our spouses and children who cleaned up around us.

References

1. Ben-Naim, A.: A Farewell to Entropy: Statistical Thermodynamics Based on Information. World Scientific, Singapore (2008)
2. Chiribella, G., D'Ariano, G.M., Perinotti, P.: Probabilistic theories with purification. Phys. Rev. A **81**(6), 062348 (2010)
3. Compton, A.H.: The Freedom of Man. Yale University Press, New Haven (1935)
4. D'Ariano, G.M., Manessi, F., Perinotti, P.: Determinism without causality. Physica Scripta, 2014(T163):014013 (2014)
5. Eddington, A.S.: The Nature of the Physical World. Cambridge University Press, Cambridge (1928)
6. Einstein, A.: Letter to Max Born, 4 December 1926. In: Born, I. (ed.) The Born-Einstein Letters. Walker and Company, New York (1971)
7. Jaynes, E.T.: Where do we stand on maximum entropy? In: Levine, R.D., Tribus, M. (eds.) The Maximum Entropy Formalism. MIT Press, Cambridge (1978)
8. Lloyd, S.: A Turing test for free will. Philos. Trans. R. Soc. A **28**, 3597–3610 (2012)
9. Moore, T.A., Schroeder, D.V.: A different approach to introducing statistical mechanics. Am. J. Phys. **65**, 25–36 (1997)
10. Schroeder, D.V.: An Introduction to Thermal Physics. Addison Wesley Longman, Reading (2000)

Chapter 13
Finding Structure in Science and Mathematics

Noson S. Yanofsky

One can view the laws of nature as having goals and intentions to produce the complex structures that we see. But there is another, deeper, way of seeing our world. The universe is full of many chaotic phenomena devoid of any goals and intents. The structure that we see comes from the amazing ability that scientists have to act like a sieve and isolate those phenomena that have certain regularities. By examining such phenomena, scientists formulate laws of nature. There is an analogous situation in mathematics in which researchers choose a subset of structures that satisfy certain axioms. In this paper, we examine the way these two processes work in tandem and show how science and mathematics progress in this way. The paper ends with a speculative note on what might be the logical conclusion of these ideas.

13.1 The Laws of Nature that We Find

Scientists look around the universe and see amazing structure. There are objects and processes of fantastic complexity. Every action in our universe follows exact laws of nature which are perfectly expressed in a mathematical language. These laws of nature are fine-tuned to bring about life, and in particular, intelligent life. It seems that the final goal of all these laws of nature is to create a creature that is in awe of the universe that created him. What exactly are these laws of nature and how do we find them?

The universe is so structured and orderly that we compare it to the most complicated and exact contraptions of the age. In the 18th and 19th century, the universe was compared to a perfectly working clock or watch. Philosophers then discussed the Watchmaker. In the 20th and 21st century, the most complicated object is a com-

N. S. Yanofsky (✉)
Brooklyn College, Brooklyn, NY, USA
e-mail: noson@sci.brooklyn.cuny.edu

© Springer International Publishing AG, part of Springer Nature 2018
A. Aguirre et al. (eds.), *Wandering Towards a Goal*, The Frontiers Collection,
https://doi.org/10.1007/978-3-319-75726-1_13

puter. The universe is compared to a perfectly working supercomputer. Researchers ask who programed this computer. The analogy is taken even further with scientists wondering if we are like characters in *The Matrix* and actually a simulation.

How does one explain all this structure? What are the goals of these laws? Why do the laws seem so perfect for producing life and why are they expressed in an exact mathematical language?

One answer to these questions is Platonism (or its cousin Realism). This is the belief that the laws of nature are objective and have always existed. They possess an exact ideal form that exists in Plato's realm. These laws are in perfect condition and they have formed the universe that we see around us. Not only do the laws of nature exist in this realm but it lives alongside all perfectly formed mathematics. This is supposed to help explain why the laws are written in the language of mathematics. Platonism leaves a lot to be desired. The main problem is that Platonism is metaphysics, not science. However, even if we were to accept it as true, many questions remain. How was this Platonic attic set up? Why does our physical universe follow these ethereal rules? How do scientists and mathematicians get access to Plato's little treasure chest of exact ideals?

The multiverse is another answer that has recently become quite fashionable (e.g. see [8, 10]). This is the belief that our universe is just one of many universes called the multiverse. Each universe has its own set of rules and its own possible structures that come along with those rules. Physicists who push the multiverse theory, believe that the laws in each universe is somewhat arbitrary. The reason why we see structure in our universe is that we happen to live in one of very few universes that have laws that can produce intelligent life. While the multiverse explains some of the structure that we see, there are questions that are left open. Rather than asking why the universe has any structure at all, we can push the question back and ask why the multiverse has any structure at all. Another problem is that while the multiverse would answer some of the questions we posed if it existed, who says it actually exists? Since we have no contact with other possible universes, the question of the existence of the multiverse is essentially metaphysics.

There is another, more interesting, explanation for the structure that is the focus of this paper. Rather than saying that the universe is very structured, say that the universe is chaotic and lacks structure. The reason why we see so much structure is that scientists act like a sieve and pull out only those phenomena that are predictable. They do not take into account all phenomena; rather, they select those phenomena they can deal with.

Some people say that science studies physical phenomena. This is simply not true. The exact shape of a cloud is a physical question that no scientist would try to describe. Who will win the next election is a physical question but no hard scientists would venture to give an absolute prediction. Whether or not a computer will halt for a given input can be seen as a physical question and yet we learned from Alan Turing that this question cannot be answered. Science does not study all physical phenomena. Rather, science studies predictable physical phenomena. It is almost a tautology: science predicts predictable phenomena.

Scientists have described the criteria for which phenomena they select: it is called *symmetry*. Symmetry is the property that despite something changing, there is some part of it that still remains the same. When you say that a face has symmetry, you mean that if the left side is swapped with the right side, it will still look the same. When physicists use the word symmetry they are discussing collections of physical phenomena. A set of phenomena has symmetry if it is the same after some change. The most obvious example is symmetry of location. This means that if one performs an experiment in two different places, the results should be the same. Symmetry of time means that the outcomes of experiments should not depend on when the experiment took place. There are many other types of symmetry [6, 9].

If phenomena are to be selected by scientists, then they must have many different types of symmetry. When a physicist sees a lot of phenomena, she must first determine if these phenomena have symmetry. She performs experiments in different places and at different times. If she achieves the same results, she then studies them to find the underlying cause. In contrast, if it failed to be symmetric, it would be ignored by the scientist.

The power of symmetry was first truly exploited by Albert Einstein. He postulated that the laws of physics should be the same even if the experimenter is moving close to the speed of light. With this symmetry requirement in mind, he was able to compose the laws of special relativity. Einstein was the first to understand that symmetry was the defining characteristic of physics. Whatever has symmetry will have a law of nature. The rest is not part of science.

A little after Einstein showed the importance of symmetry for the scientific endeavor, Emmy Noether proved an important theorem that established a connection between symmetry and conservation laws. Again, if there is symmetry, then there will be conservation laws. The physicists must be a sieve and allow the phenomena that do not possess symmetry to slip through her fingers.

There is actually something deeper going on here. The laws of physics cannot be found without "bracketing out" different phenomena. Consider the way the laws of physics are given as they are taught in any physics class. While they are not exactly false, they are totally useless! (See [1, 2, 7]). Ponder the simple law that Newton taught us about gravity. The force between two objects is given by the product of the two masses divided by the square of the distance. That is,

$$F = G \frac{m_1 m_2}{d^2}$$

This is only useful when the two bodies are spherically symmetric. The bodies also have to be totally homogenous (the mass must be evenly distributed). There cannot be any third body or gravitational forces affecting either of the bodies. Both bodies must be magnetically and electrically neutral. They cannot be traveling near the speed of light (lest the theory of special relativity take over). The bodies also cannot be too small (lest quantum effects come into play). And the list goes on. In summation, it is safe to say that there were probably never two bodies that exactly satisfied the requirements for Newton's laws to be exactly true. Rather, the physicists must make controlled experiments and ignore all these other effects. By selecting the phenomena, he idealizes the actual world to find the ideal laws of nature. Without selecting the phenomena, no such laws can be found.

No one is asserting that selecting subsets of phenomena is the only way of finding laws of nature. There are other methods for finding such laws. For example, in statistical mechanics and in quantum theory, one considers large ensembles of phenomena to be one phenomenon (a quotient set of phenomena, rather than a subset). While we acknowledge the existence of other methods, in this paper we will focus on our selection method.

There are a few problems with this explanation of the structure found in the universe. For one, it seems that phenomena that we do select and that have laws of nature are exactly the phenomena that generate *all* the phenomena. So, while the shape of a cloud or the winner of an election are too complicated for the scientist to worry about, they are generated by laws of water molecules and brain synapses that are part of science. Where is the boundary between science and non-science?

Despite these failings of our explanation for the structure, we believe it is the best candidate for being *the* solution. It is one of the only solutions that does not invoke any metaphysical principle or the existence of a multitude of unseen universes. We do not have to look outside the universe to find a cause for the structure that we find in the universe. Rather, we look at how we are looking at phenomena.

Before we move on, we should point out that our solution has a property in common with the multiverse solution. We postulated that, for the most part, the universe is chaotic and there is not so much structure in it. We, however, focus only on the small amount of structure that there is. Similarly, one who believes in the multiverse believes that most of the multiverse lacks structure [8, 10]. It is only in a select few universes that we do find any structure. And we inhabitants of this structured universe are focused on that rare structure. Both solutions are about focusing on the small amount of structure in a chaotic whole.

13.2   A Hierarchy of Number Systems

This idea that we only see structure because we are focusing on a subset of phenomena is novel and hard to wrap one's head around. There is an analogous situation in mathematics that is much easier to understand. We will focus on one important example where this selection process is seen very clearly. First we need to take a little tour of several number systems and their properties.

Consider the real numbers. In the beginning of high school, the teacher draws the real number line on the board and says that these are all the numbers one will ever need. Given two real numbers, we know how to add, subtract, multiply and divide them. They comprise a number system that is used in every aspect of science. The real numbers also have an important property: they are totally ordered. That means that given any two different real numbers, one is less than the other. Just think of the real number line: given any two different points on the line, one will be to the right of the other. This property is so obvious that it is barely mentioned.

While the real numbers seem like a complete picture, the story does not end there. Already in the 16th century, mathematicians started looking at more complicated number systems. They began working with an "imaginary" number i which has the property that when it is squared it is -1. This is in stark contrast to any real number whose square is never negative. They defined an imaginary number as the product of a real number and i. Mathematicians went on to define a complex number that is the sum of a real number and an imaginary number. If r_1 and r_2 are real numbers, then $r_1 + r_2 i$ is a complex number. Since a complex number is built from two real numbers and we usually draw them in a two-dimensional plane. The real number line sits in the complex plane. This corresponds to the fact that every real number, r_1, can be seen as the complex number $r_1 + 0i$.

We know how to add, subtract, multiply, and divide complex numbers. However, there is one property that is different about the complex numbers. In contrast to the real numbers, the complex numbers are not totally ordered. Given two complex numbers, say $3 + 7.2i$ and $6 - 4i$, can we tell which one is more and which one is less? There is no obvious answer. (In fact, one can totally order the complex numbers but the ordering will not respect the multiplication of complex numbers). The fact that the complex numbers are not totally ordered means that we lose structure when we go from the real numbers to the complex numbers.

The story is not over with the complex numbers. Just as one can construct the complex numbers from pairs of real numbers, so too, one can construct the quaternions from pairs of complex numbers. Let $c_1 = r_1 + r_2 i$ and $c_2 = r_3 + r_4 i$ be complex numbers; then we can construct a quaternion as $q = c_1 + c_2 j$ where j is a special number. It turns out that every quaternion can be written as

$$r_1 + r_2 i + r_3 j + r_4 k,$$

where i, j, and k are special numbers. So while the complex numbers are comprised of two real numbers, the quaternions are comprised of four real numbers. Every complex number $r_1 + r_2 i$ can be seen as a special type of quaternion: $r_1 + r_2 i + 0j + 0k$. We can think of the quaternions as a four-dimensional space which has the complex numbers as a two-dimensional subset of it. We humans have a hard time visualizing such higher-dimensional spaces.

The quaternions are a full-fledged number system. They can be added, subtracted, multiplied and divided with ease. Like the complex numbers, they fail to be totally ordered. But they have even less structure than the complex numbers. While the multiplication of complex numbers is commutative, that is, for all complex numbers c_1 and c_2 we have that $c_1 c_2 = c_2 c_1$, this is not true for all quaternions. This means there are quaternions q_1 and q_2 such that $q_1 q_2$ is different than $q_2 q_1$.

This process of doubling a number system is called the "Cayley–Dickson construction," named after the mathematicians Arthur Cayley and Leonard Eugene Dickson. Given a certain type of number system, one gets another number system that is twice the dimension of the original system. The new system that one develops has less structure (i.e. fewer axioms) than the starting system.

If we apply the Cayley–Dickson construction to the quaternions, we get the number system called the *octonions*. (See e.g. [4]). This is an eight-dimensional number system. That means that each of the octonions can be written with eight real numbers as

$$r_1 + r_2\mathbf{i} + r_3\mathbf{j} + r_4\mathbf{k} + r_5\mathbf{l} + r_6\mathbf{m} + r_7\mathbf{n} + r_8\mathbf{p}$$

Although it is slightly complicated, we know how to add, subtract, multiply, and divide octonions. Every quaternion can be written as a special type of octonion in which the last four coefficients are zero.

Like the quaternions, the octonions are neither totally ordered nor commutative. However, the octonions also fail to be associative. In detail, all the number systems that we have so far discussed possess the associative property. This means that for any three elements, a, b, and c, the two ways of multiplying them, a(bc) and (ab)c, are equal. However, associativity does not always work with the octonions. That is, there exists octonions o_1, o_2 and o_3 such that $o_1 (o_2 o_3) \neq (o_1 o_2) o_3$.

We can go on with this doubling and get an even larger, sixteen-dimensional number system called the *sedenions*. In order to describe a sedonian, one would have to give sixteen real numbers. Octonions are a special type of sedonian: their last eight coefficients are all zero. But researchers steer clear of sedenions because they lose an important property. While one can add, subtract, and multiply sedenions, there is no way to divide them nicely. Most physicists think this is beyond the pale and "just" mathematics. Even mathematicians find sedenions hard to deal with. One can go on to formulate 32-dimensional number systems and 64-dimensional number systems, etc. But they are usually not discussed because, as of now, they do not have many applications. We will concentrate on the octonions.

Here is a diagram of all these different number systems and the properties they have:

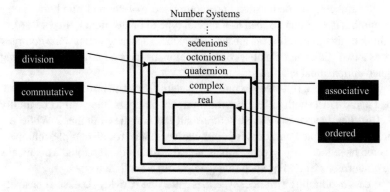

Let us discuss the applicability of these number systems. The real numbers are used in every aspect of physics. All quantities, measurements, durations, and lengths of physical objects are given as real numbers. Although complex numbers were formulated by mathematicians to help solve equations ($\mathbf{i}$ is the solution to the equation $x^2 = -1$), in the middle of the 19th century, physicists started using complex numbers to discuss waves. In the 20th century, complex numbers became fundamental for the study of quantum mechanics. By now, the role of complex numbers is very important in many different branches of physics. The quaternions show up in physics but are not a major player. The octonions, the sedenions, and the larger number systems rarely arise in the physics literature.

13.3 The Laws of Mathematics that We Find

The usual view of these number systems is to think that the real numbers are fundamental while the complex, quaternions, and octonions are strange larger sets that keep mathematicians and some physicists busy. The larger number systems seem unimportant and less interesting.

Let us turn this view on its head. Rather than looking at the real numbers as central and the octonions as strange larger number systems, think of the octonions as fundamental and all the other number systems as just special subsets of octonions. The only number system that really exists is the octonions. To paraphrase Leopold Kronecker, "God made the octonions, all else is the work of man." The octonions contain every number that we will ever need. (And, as we stated earlier, we can do the same trick with the sedenions and even the 64-dimensional number system. We shall fix our ideas with the octonions).

Let us explore how we can derive all the properties of the number systems that we are familiar with. Although the multiplication in the octonions is not associative, if one wants an associative multiplication, one can look at a special subset of the octonions. (We are using the word "subset" but we need a special type of subset that respects the operations of the number system. Such subsets are called "subgroups," "subfields," or "sub-normed-division-algebras," etc. For the reader's benefit, we use "subset"). So if one selects the subset of all octonions of the form

$$r_1 + r_2\mathbf{i} + r_3\mathbf{j} + r_4\mathbf{k} + 0\mathbf{l} + 0\mathbf{m} + 0\mathbf{n} + 0\mathbf{p},$$

then the multiplication will be associative (like the quaternions). If one further looks at all the octonions of the form

$$r_1 + r_2\mathbf{i} + 0\mathbf{j} + 0\mathbf{k} + 0\mathbf{l} + 0\mathbf{m} + 0\mathbf{n} + 0\mathbf{p},$$

then the multiplication will be commutative (like the complex numbers). If one further selects all the octonions of the form

$$r_1 + 0\mathbf{i} + 0\mathbf{j} + 0\mathbf{k} + 0\mathbf{l} + 0\mathbf{m} + 0\mathbf{n} + 0\mathbf{p},$$

then they will have a totally ordered number system. All the axioms that one wants satisfied are found "sitting inside" the octonions.

This is not strange. Whenever we have a structure, we can focus on a subset of special elements that satisfies certain properties. Take, for example, any group. We can go through the elements of the group and pick out those X such that, for all elements Y, we have that $XY = YX$. This subset is a commutative (abelian) group. That is, it is a fact that in any group there is a subset which is a commutative group. We simply select those parts that satisfy the axiom and ignore ("bracket out") those that do not. The point we are making is that if a system has a certain structure, special subsets of that system will satisfy more axioms than the starting system.

This is similar to what we are doing in physics. We do not look at all phenomena. Rather, we pick out those phenomena that satisfy the requirements of symmetry and predictability. In mathematics, we describe the subset with the axiom that describes it. In physics, we describe the selected subset of phenomena with a law of nature.

13.4   Working in Tandem and Moving Forward

We have shown that there is an important analogy between physics and mathematics. In both fields, if we do not look at the entirety of a system, but rather look at special subsets of the system, we see more structure. In physics we select certain phenomena (the ones that have a type of symmetry) and ignore the rest. In mathematics we are looking at certain subsets of structures and ignore the rest. These two bracketing operations work hand in hand.

The job of physics is to describe a function from the collection of observed physical phenomena to mathematical structure:

$$\text{observed physical phenomena} \rightarrow \text{mathematical structure}.$$

That is, we have to give mathematical structure to the world we observe. As physics advances and we try to understand more and more observed physical phenomena, we need larger and larger classes of mathematics. In terms of this function, if we are to enlarge the input of the function, we need to enlarge the output of the function.

Some examples of this broadening of physics and mathematics are needed. (i) When physicists started working with quantum mechanics they realized that the totally ordered real numbers are too restrictive for their needs. They required a number system with fewer axioms. They found the complex numbers. (ii) When Albert Einstein wanted to describe general relativity, he realized that the mathematical structure of Euclidean space with its axiom of flatness (Euclid's fifth axiom) was too restrictive. He needed curved, non-Euclidian space to describe the space-time of general relativity. (iii) In quantum mechanics it is known that for some systems, if we first measure X and then Y, we will get different results than first measuring Y and then measuring X. In order to describe this situation mathematically, one needed to leave the nice world of commutativity. They required the larger class of structures where commutativity is not assumed. (iv) When Boltzmann and Gibbs started talking about statistical mechanics, they realized that the laws they were coming up with were no longer deterministic. Outcomes of experiments no longer either happen (p $(X) = 1$) or do not happen (p $(X) = 0$). Rather, with statistical mechanics one needs probability theory. The chance of a certain outcome of an experiment is a probability $(p(X))$ is an element of the infinite set $[0, 1]$ rather than the restrictive finite subset $\{0, 1\}$). (v) When scientists started talking about the logic of quantum events, they realized that the usual logic, which is distributive, is too restrictive. They needed to formulate the larger class of logics in which the distributive axiom does not necessarily hold true. This is now called quantum logic. Many other examples exist.

Paul A.M. Dirac understood this loosening of axioms about 85 years ago when he wrote the following:

> The steady progress of physics requires for its theoretical formulation a mathematics which get continually more advanced. This is only natural and to be expected. What however was not expected by the scientific workers of the last century was the particular form that the line of advancement of mathematics would take, namely it was expected that mathematics would get more and more complicated, but would rest on a permanent basis of axioms and definitions, while actually the modern physical developments have required a mathematics that continually shifts its foundation and gets more abstract. Non- euclidean geometry and noncommutative algebra, which were at one time were considered to be purely fictions of the mind and pastimes of logical thinkers, have now been found to be very necessary for the description of general facts of the physical world. It seems likely that this process of increasing abstraction will continue in the future and the advance in physics is to be associated with continual modification and generalisation of the axioms at the base of mathematics rather than with a logical development of any one mathematical scheme on a fixed foundation [3].

As physics progresses and we become aware of more and more physical phenomena, larger and larger classes of mathematical structures are needed and we get them by looking at fewer and fewer axioms. There is no doubt that if Dirac lived now, he would talk about the rise of octonions and even the sedenions within the needed number systems.

In order to describe more phenomena, we will need larger and larger classes of mathematical structures and hence fewer and fewer axioms. What is the logical conclusion to this trend? How far can this go? Physics wants to describe more and more phenomena in our universe. Let us say we were interested in describing *all* phenomena in our universe. That is, we don't "bracket out" any phenomena. What type of mathematics would we need? How many axioms would be needed for the mathematical structure to describe all the phenomena? Of course, it is hard to predict but it is even harder not to speculate. One possible conclusion would be that if we look at the universe in totality and not bracket any subset of phenomena, the mathematics we would need would have no axioms at all. That is, the universe in totality is devoid of structure and needs no axioms. Total Lawlessness! This would finally eliminate all metaphysics when dealing with the laws of nature and mathematical structure. It is only the way that we look at the universe that gives us the illusion of structure.

With this view of physics, we come to even more profound questions. These are the future projects of science. If the structure that we see is only an illusion, then why do we see this illusion? Instead of looking at the laws of nature that are formulated by scientists, we have to look at scientists and the way they pick out subsets of phenomena and their concomitant laws of nature. What is it about human beings that renders us so good at being sieves? Rather than looking at the universe, we should look at the *way* we look at the universe [5, 11].

Acknowledgements I am grateful to Jim Cox, Avi Rabinowitz, Karen Kletter, Karl Svozil, and Mark Zelcer for many helpful conversations.

References

1. Cartwright, N.: How the Laws of Physics Lie. Oxford University Press, Oxford (1983)
2. Cartwright, N.: The Dappled World: A Study of the Boundaries of Science. Cambridge University Press, Cambridge (1999)
3. Dirac, P.A.M.: Quantised singularities in the electromagnetic field. Proc. Roy. Soc. **133A**, 60–72 (1931)
4. Dray, T., Manogue, C.A.: The Geometry of the Octonions. World Scientific Publishing Company, Singapore (2015)
5. Eddington, A.S.: Philosophy of Physical Science. Cambridge University Press, Cambridge (1939)
6. Van Fraassen, B.C.: Laws and Symmetry. Oxford University Press, Oxford (1989)
7. Giere, R.N.: Science without Laws. University Of Chicago Press, Chicago (1999)
8. Greene, B.: The Hidden Reality: Parallel Universes and the Deep Laws of the Cosmos. Knopf, New York (2011)
9. Stenger, V.J.: The Comprehensible Cosmos: Where Do the Laws of Physics Come From? Prometheus Books, USA (2006)
10. Tegmark, M.: Our Mathematical Universe: My Quest for the Ultimate Nature of Reality. Knopf, New York (2014)
11. Yanofsky, N.S.: The Outer Limits of Reason: What Science, Mathematics, and Logic Cannot Tell Us. MIT Press, Cambridge (2013)

Chapter 14
From Athena to AI: The Past and Future of Intention in Nature

Rick Searle

The fault, dear Brutus, is not in our stars, But in ourselves, that we are underlings.

William Shakespeare, Julius Caesar

14.1 Prologue: Gods, Water, Fire

So it was that Aristagoras of Miletus the great defender of the gods conspired to bring together all those Milesians and near Hellenes who were in those days beginning to spread doubts regarding the power and providence of the gods to control the fates of mortals and the happenings of the world.

Inspired by the goddess Athena, and paid for by his enormous wealth, Aristagoras called a for a great banquet of all the nobles of Miletus and the surrounding lands where the philosophers would engage in a sort of wrestling match of intellects over the question of how mindless matter could ever give rise to intention and aims. The hope of Aristagoras being not that one philosopher would emerge the clear victor from such a contest, but that each would so contradict and exhaust the others that in the end he could close the banquet with the affirmation that the ancient faith was in fact correct, that matter lacking mind could do nothing, and that all that happened ever was and ever will be a reflection of the intention of either gods or mortals.

The thinkers who attended Aristagoras' banquet were among the most prominent of the day. There was Thales the water worshiper, Parmenides and his guard dog Zeno, the senile Pythagoras and his noble heir Philocrates. In addition there was the laughing philosopher Democritus and Heraclitus the never- wet. Never before or since has there been such a meeting of the world's greatest minds.

R. Searle (✉)
Institute of Ethics and Emerging Technologies, Harrisburg, US
e-mail: rsearle.searle@gmail.com

© Springer International Publishing AG, part of Springer Nature 2018
A. Aguirre et al. (eds.), *Wandering Towards a Goal*, The Frontiers Collection,
https://doi.org/10.1007/978-3-319-75726-1_14

For dramatic effect Aristagoras had arranged it that his questions were spoken by three women dressed in black veils like the goddesses of fate seen in plays. It was Thales to whom Aristagoras put the first question.

"Thales son of Asherah", the chorus sang, "you say all the world is made of water, but is not the world full of gods, suffused with intention and will? How then can mindless matter act so?" Thales rose, raised his cup of wine to his host Aristagoras, took a sip to wet his throat and began.

"Surely many of you think that each great river is under the sovereignty of a god. Some think similarly of even small streams of which they are intimate and that their every ebb and eddy is under the providence of some lesser spirit. Imagine now that one were an ant standing on the table before me" Thales gently placed his finger on the table in front of him. He poured a small trickle of wine from his cup onto the table and smiling continued "would one not conclude, if positing that all bodies of water had their own gods that such a trickle as this had its own god assigned to it? Are all things, even the very smallest, thus full of gods? Would these gods be more than gnats to Lord Poseidon ruling over the oceans? And even the largest bodies of water on earth would be but droplets to the great God looking over all the cosmos. Are not our own bodies made out of this same water? So if the water of which our bodies is made is without mind, then water can give rise to mind and its aims and intentions, even if I cannot say how this is so." At this Thales sat back down. The chorus then turned to Parmenides and sang:

"Arise, wisest of Elea, to answer our question and respond to song of Thales." Parmenides arose while his loyal Zeno sat beside him twiddling with a ball of yarn like Daedalus and began:

"I believe that nature is ruled by Thales' great God much more so than the fiction of poets like Hesiod and Homer. If there are gods, they are above the passions of men. And if the god's are subject to reason is there not one reason to rule them all? If all that exists is made of one substance, as Thales claims, then all the distinctions between one thing and another, between the world and ourselves must be a mirage. It is in light of this that I maintain, despite our senses, that all change is an illusion. That what will happen has been established from eternity in such a way that we can understand it to have already happened. To an ant Thales might be thought a god controlling a river of red wine, but Thales himself had been fated to create his river in such a way since eternity."

At this there was a great murmuring of disbelief and even laughter from the crowd upon which Zeno stood up and began speaking in one of his riddles.

"If what we call the past is the cause of the present, and our current present is the past of what we call the future, then everything has not only been determined for us, in our time, it has been determined for the future as well. One with a complete knowledge of the conditions in which an arrow was shot from a bow would also have complete knowledge of its entire flight and know for certain, and beforehand, where it would ultimately strike. One who had complete knowledge of how an arrow had struck its target would likewise know everything about its flight back to the time it left the bow. Yet why privilege beginnings and endings? Complete knowledge of any point gives complete knowledge of past and future. We mortals are the arrows shot from the bow of God."

The chorus heckled: "Then to our hearts what you say be well. Puppets you claim us to be, and yet not to fate but to some greater power greater still?" At which Parmenides stood up and responded.

"As with your namesake fate what is… is… what is not… is not- but with this difference- no gods inspiring the lechery of Paris lie behind the destruction of Troy, nor even the will of the justice seeking Achaeans, but the logos of the kosmos itself made it so. Mortals believe themselves free but are no freer to choose what will be or what will not than characters in the mind of Homer. Aims and intentions are mere words, an illusion."

"And the gods?" The chorus asked?

"In a universe where nothing happens no gods are needed," Parmenides answered. We call God all that exists in such an eternity. Everything that is exists as an eternal unchanging thought in the mind of the great and only God."

Accusations of impiety rolled through the crowd.

"Free us then, oh wise Pythagoras, from the Elean's impiety that melds all the great gods of Olympus and mortals into a single mind" sang the chorus. At their words Pythagoras did not stand or speak, but pulled out from behind where he sat a golden lyre with a single string. He plucked the string which made a sort of low bellow and then moved his finger to the halfway point and plucked it again. He continued along like this dividing the string of the lyre at various points and plucking the string. After sometime Aristagoras became annoyed at Pythagoras' wordless performance and stood up ready to dismiss the old crow whose mind had clearly rotted with age when his loyal friend the young Philocrates stood up and spoke for his friend and teacher.

"Can you not hear? Are you not amazed that a string cut in half plays the same note only higher? Can you not hear that some divisions of the string played one after the other sound beautiful to the ear while others induce pain? This is the great discovery which our Pythian Apollo has brought into the world. That the logos of which Parmenides speaks is composed in the language of music and number which we can hear and understand."

Philocrates looked down at Pythagoras who remained seated, and the latter nodded in agreement at which Philocrates continued. "This is the mind of God of which mortals can partake. Do you not see this order when you look to the night darkened heavens and see its glory and regular motions, or when you stare intently at the beauty of your lover? All living things that possess a *nous,* a mind, can see this kosmos, this order and beauty of the world, but among mortals only humans can *understand* this order by grasping its logos."

Carried away by the force of his logic Philocrates went on. "This knowledge is not for contemplation alone, for understanding the logos leads to enormous power. Once a player on the lyre of the kosmos knows its rules he can use them to achieve his ends, to imagine and play a tune of his own. Minds then are not fated as the gloom laden Parmenides claims, and mind, comprehending the logos itself makes this so."

At this the chorus sang "So Thales says the world is water and that mind arises from such, though he cannot say not how or when. And Parmenides and his guard dog think there is just one mind that never changes, and in thinking otherwise our minds themselves are fooled, whereas Pythagoras and Philocrates think all are notes and number and that that while nature is ordered in knowing its rules we can direct the play. Now to you, the most notorious of atheists", the chorus said turning to Democritus. "Tell us how aims and intention can arise from mindless matter, and answer those who have spoken."

Democritus arose from his seat clearly amused and let out a belch. It appeared he may have already been a little drunk.

"See here how my noble friends are only half correct, for such is the danger of not drinking enough when the wine is free. Thales claims all the world is made of one substance and that this substance is water. Right he is to look for one such element from which the world is built, but surely water is merely a metaphor. For how can one make fire out of water? We must go below."

"Parmenides agrees that all is composed of one underlying substance, but then imagines a blob larger than his bulbous head. And this is the problem as well with Zeno and his arrow, not the size of his noggin but what is in it and comes out of its mouth, for as I have heard elsewhere with his tale of Achilles and the Tortoise, Zeno believes that number is infinitely divisible. Yet numbers are just playthings of the mind with which we can do things impossible in reality. No real matter can be divided so. Take a grain of sand and divide it and eventually one will reach the prime element. I call these elements *atoms*." Democritus chugged down his goblet of wine and gestured to a servant that it be refilled. He continued:

"My friends, Pythagoras and Philocrates are obsessed with the toys of geometry and though they are right see the order which mathematical arrangements of matter can bring, they fail to see that most arrangements are meaningless except to those who claim to interpret them. Chaos rules the world and if we find some arrangement of atoms beautiful it is only because they accidentally match our own like a flower that reminds one of the face of his beloved. Aims and intentions emerge because some accidental ordered arrangements- minds- require other accidental arrangements- like good food and wine- to live and thrive. There are no gods or great God required, only atoms and the eternal flux of time."

The crowd gasped at which Democritus let out another belch. Aristagoras gestured that the laughing philosopher should be seated and the chorus sang pleadingly to the only philosopher yet to speak:

"It is up to you, oh bedeviling Heraclitus, to save the gods and the freedom of mortals from the philosophers' impiety" the chorus plaintively sang. At this Heraclitus stood up and began:

"Save you I cannot, for I agree with much of what my friends have said, and hope here only to help buttress the ship of their thought so that it is more worthy of the rough seas the future will inevitably bring. I mostly agree with the interpretation of laughing Democritus and even his views on the words of our friends. What I would add is that those things which preserve their identity in the midst of his sea of randomness must do so by preserving their patterns much more so than retaining their individual elements which he calls atoms."

"A stream is such a pattern of ever changing atoms which preserve its shape, as a fire can if tended well. And what are minds but the tending of a fire by itself, making sure it is fed neither too much nor too little?"

"Minds are patterns meant to uncover other patterns in the world around them. Yet such patterns exists not merely in the mind but are born out of our encounter with nature, the same nature other creatures encounter and in which they can discover patterns similar to our own. It has been said by one of our geometers that the honeybee has discovered the best way to divide an area into equal sections with the least possible perimeter?[1] Minds are nature's way of discovering its own patterns.

"Humans share this trait with all of the living, though there is a distinction. Animals discover and act on patterns mostly without thinking, but humans reason, we think and plan, which saves us almost uniquely from being frozen in the block of time Parmenides imagined and frees us, as Philocrates said, to understand the logos of nature so as to compose realities of our own making."

"For now, we have no need of gods. Aims and intentions can arise from mindless matter which is the consequence of atoms driven to preserve, comprehend, and even create, their own patterns. Where human power in such regard ends who can say? Perhaps someday far into the future mortals will manage to infuse the pattern of mind into matter itself like the automata of Hephaestus imagined by Homer. In that case the world full of gods which our ancestors believed, and most of you still believe in, will actually be our own." At this Heraclitus bowed to the crowd and sat down.

Aristagoras felt himself dumbfounded by Heraclitus' words, and unanchored by what had transpired. Instead of ending the banquet with a ringing defense of the gods as he had intended, he ordered the customary libations for a voyage, the sacrifice of a bull to Athena goddess of wisdom, pouring wine as an offering, and intoning a prayer for future safety. Aristagoras could not understand where the philosophers were taking us, only that all mortals had been impressed into the journey.[2]

14.2 How Aims and Intention Arise from Mindless Matter

It may seem strange to have started an essay which hopes to address the question of how mindless mathematical laws can give rise to aims and intentions with an imagined dialogue of philosophers from over 2,500 years ago, but I had very good reasons for doing so. It was from these pre-Socratic philosophers that we can trace our own scientific worldview. For despite the apparent naiveté of their various theories, what the pre-Socratics were the first to do was to seek out explanations for the behavior and order of nature that were independent of the will of the gods. As the former soldier and author Roy Scranton writes of civilization before this first ancient enlightenment in his dark meditation on civilization's fate, *Learning to die in the Anthropocene*:

> Yet as humans evolved complex social networks, language, consciousness, and then culture, we came to organize ourselves through systems that saw not merely agency in the world, but will.

When Homer's Greeks stalked the battlefield, Ares drove them in frenzies to kill and Athena stayed their hands. For those ancients, the will of men was subject to the will of the gods and all were ruled by fate. Causality was comprehended by seeing the universe as a web of personified forces. It was only later, after the rise of literacy that Greek poets and sages began to articulate a difference we take as fundamental today, the distinction between human will and natural force. The independent persistence of written language-logos- became the structuring metaphor for the independent persistence of the human mind. We began to believe in the freedom of thought.[3]

In closing the door to the gods, the pre-Socratics not only managed to free us from the mistaken belief that even lifeless matter acted with aims and intentions towards *us*- but- and in the opposite direction- opened up a realm of freedom by dispensing with fate and the ill intent of the natural world itself. If we could only understand the rules by which nature operated we could leverage and adjust to those rules to obtain our own ends, or at the very least, obtain some degree of safety.

In an admittedly very simplified reading, the whole history of science, along with the technology that flows from science, has been the story of the discovery, loss, and recovery of the idea that the world is not ruled by will. In the 1600's thinkers finally started to move away from a model of the world where every flower was opened via the will of God. Yet this move away from will came to be dependent on an image of God as clockmaker with all of time unfolding along the kind of deterministic course that could be seen in thinkers as diverse as Spinoza, Leibniz, and Calvin reaching a peak of elegance and unprecedented scientific importance with the publication of Newton's *Principia Mathematica*.[4]

The deterministic worldview present in these thinkers would have warmed Parmenides' heart and modern physics' interpretation of God as the world's mathematician would have seemed the most noble of legacies to Pythagoras and his followers, as in some sense they actually are.

Yet such laws have their explanatory limits in that it is in the very noise they compress and smooth away (the diversity and granularity of experience) that some of the most essential information for action, the field of aims and intentions, lies. Darwin, with clear echoes of Democritus, gave us a way in which a single input could result in multiple possible outcomes. In the smallest of changes lie innumerable possibilities.[5]

It has taken a very long time since Darwin published his *Origin of Species* for evolutionary thinking to become sophisticated enough to inform the deterministic branches of science, and to gain the quantitative depth necessary to be able to borrow from, and engage in a dialogue of equals with, more mathematically based sciences such as physics. Though in its early days, what this conversation has shown so far is something close to that of the view of my imagined Heraclitus.

As we all known from the Second Law of Thermodynamics, all large scale structures in the universe can survive overtime only if as a consequence of their order they displace an equal amount of disorder in the form of heat. What Heraclitus had over Democritus was his recognition that every ordered system is in a race between its own efforts to preserve its structure and those forces aiming to pull it apart. And Heraclitus didn't just think this struggle against chaos was something done by living things alone. He thought a river, and especially a phenomenon like fire, shared these features as well.

Nowadays, biophysicists, most notably Jeremy England of MIT, have pointed out how adaptation- meaning the efficient absorption of energy from a fluctuating environment- arises in any (even non-living) system able to displace excess entropy into a surrounding bath. Structure and organized behavior flow almost inevitably from physics itself. What life adds to this equation isn't so much adaptation as the extensions of lessons from such adaptations into the future through reproduction.[6] This move towards greater complexity (not to be confused with 19th century notions of "progress") isn't just an accidental byproduct of natural selection but is built into the very nature of both biological[7] and human-made systems.[8]

This kind of erosion of the philosophical boundaries between the living and the nonliving when it comes to lifelike behaviors which England represents is taking place at the level of cognition as well. While something like Giulio Tonoi's concept of Integrated Information Theory may ultimately prove wanting as a theory of consciousness, it does point us in the direction of a reality where consciousness, and therefore the aims and intentions that come with consciousness, emerges as a natural consequence of systems integrated in a peculiar way independent of the substrate in which those processes occur. Mind has its origins not just in matter, but in matter organized in a very specific way.[9]

We now know that a network of complex interactions gives rise to mind (what the ancient Greeks called nous) in the brain. We also now know that both intelligence along with the aims and intentions of agency, not only arise from different forms of brains such as the distributed nervous system of cephalopods,[10] or using a completely different form of neuro-physiology, such as ctenophores,[11] but as is suggested in the work of Tonoi, something like mind and intention arise in systems that do not possess a brain or nervous system at all.

Much of the natural world is filled with these non-brain based minds that were formerly hidden from us. What are perhaps commonly perceived to be largely mindless, individualistic organisms, namely trees, are now known to be deeply interconnected via mycorrhizal fungi[12] to create the so-called "wood-wide web" a system through which trees perform such apparently mindful and intentional activities as warning of danger and sharing resources.[13]

Nature abounds with phenomenon such as swarms where the action of individuals tied together into some mutually reinforcing systems seems to give rise to collective behavior that appears mindful (at least in the sense of being able to perceive the world around them) and intentional.

The beginnings of complexity would thus seem to have an easy entry way, given the laws of physics. However, what allows the complexity of such adaptive systems to grow across time is the fact that they are the product of very unique histories. The vast majority of human aims and intentions have their roots in either evolutionary history (conveyed by genes and epigenetic changes), or the accumulated past of our cultures, and, of course, grown out of our own memory of deeply personal experiences.

In a sense the fact that we have aims and intentions at all is a consequence of the fact that anything like Parmenides' view of a timeless block-like universe remains out of reach for us in anything more than an imaginative or abstract sense. Our intelligence and decision making, indeed our very experience of freedom itself, emerges in this tension between thought and action, the gap between our internal models of the world and reality itself. We are tuned by the outside world and tune ourselves to the world we are enmeshed in like a Pythagorean lyre. As the technologist Jaron Lanier has put it:

> The cybernetic structure of a person has been refined by a very large, very long, and very deep encounter with physical reality.[14]

Even our most intelligent machines are nowhere near us in this form of evolved complexity of their mental structures and consequent behaviors, and perhaps they never will be. Yet those who have made the greatest contributions to artificial intelligence so far are those who have embraced evolutionary techniques, which at the very least gives us a cartoon like replay of how human intelligence must have emerged.[15]

Indeed it has been shown that even simple computer programs can exhibit complex structure and behavior through having to adapt to a world of simple rules, as if Pythagoras's lyre could discover its own tune by being rewarded every time it stumbled across combinations of notes that were pleasing to the ear.[16] This leaves us with a question: if such programs could be asked why they took the action they did they might respond as if they had some choice in the matter rather than being driven, as they certainly are, by their underlying algorithm and its history, which leads us to wonder whether our own aims and intentions might likewise be illusions, mere white- noise emanating from an underlying program?

Philosophers have given us a way to avoid this conclusion and in ways that dovetail nicely with arguments for the natural emergence of complex systems in a universe moving in the direction of increased entropy.

What sets complex adaptive systems apart from other types of systems is their ability to respond not just too external cues from their environment, but to signals emanating from within the system itself. The philosopher Daniel Dennett has shown how real freedom, or as he calls it "freedom worth wanting", could evolve even in a completely deterministic universe because the very point of evolving consciousness in the first place is for organisms to decide between alternative aims and intentions, and that it is this ability to decide between options that constitutes our freedom. Consciousness is thus tied to freedom even if its possession is unnecessary for simple agency, that is, for an entity to be able to act.

Higher animals therefore possess some degree of freedom, yet human beings are uniquely free because only we among them can use our imagination and language to constrain or guide our behavior against the impact of our immediate environment or even our evolutionary history to expand our range of choices. Until the forms of artificial intelligence we create can likewise reason about their decisions and leverage their internal states (their equivalents of imagination and emotion) they will remain mere tools.[17]

14.3 Conclusion: Hyperobjects and the Retreat of Logos

Until machines sharing similar features with life and the human mind are created, or until we discover similar forms of intelligence beyond the earth, our cosmological status in terms of our freedom and therefore moral responsibility will remain unique. This freedom is a consequence of our acquired knowledge as much as biology and therefore is the gift of our history, which, beginning with the pre-Socratics, has given us a more extensive capacity for action than has ever been experienced before, perhaps anywhere in the universe. Strangely, this very history seems to be leading us into a world where not only our newly achieved freedom but our understanding, what the Greeks called *logos*, are in danger of being lost.

The way in which we have arrived at our unprecedented degree of freedom, over and above Dennett's sense of it, and even while continuing to assume we live in a largely deterministic universe, has been pointed out by the philosopher Jenna Ishmael. Understanding causality allows us to conceptually extract elements from nature subjectable to our influence in order to change them and therefore bring some sort of deterministic outcome closer to our desired end.[18] Action at the causal level is an attempt to tip the scales in favor of some possibility we find desirable, which takes advantage of the fact that in nature, as William James put it, "The parts have a certain loose play upon one another..."[19]

It is this degree of freedom over nature that has been the backdrop in which our political freedom was won. Whereas what we see in nature may, as Scranton wrote, have agency but not will, human beings, in the eyes of Dennett, Ishmael, and James really do have will, and even a very real type of freedom. We can intend for some future to occur, even one quite far off in the future, and to a limited extent at least, can cause it to happen. We possess this freedom to some extent as individuals, but to a much greater degree, and more importantly, collectively.

This collective aspect to our freedom is too often missed.[20] From its very beginnings science has been a collaborative project, and this is the case even when individual scientists are driven by intense rivalries. The astounding progress of science and technology in the modern era has been largely driven by the improved ability of people to communicate with one another all of whom were driven by a common project.[21] That shared end was to better understand nature and more effectively act upon it. In terms the ancient Greeks might understand the goal was to discover the logos (the reason) behind the intuitively perceived kosmos (the order and beauty) of the world.

Yet if such a knowledge of the world's order was achieved collectively this understanding itself was still the possession of the individual, and, in its early days, it seemed that every educated individual might someday too be able to fully grasp this order. In the 18th and 19th centuries it seemed reasonable to assume that not only would humanity as a species possess a complete knowledge of nature's laws, but that such knowledge would be understood by individuals and shared by an ever larger portion of the non-scientist public as well.[22]

What is disorienting is that in the 21st century we have come to a place where both mind and its agency are understood to emerge from the "bottom up", from complex interactions between distinct elements which are then combined with often externalized forms of memory, but simultaneously seemed to have lost any place in which individual human agency and logos- meaning freedom and comprehension- might permanently stand.

Human made systems, themselves enabled by the scientific revolution, have taken on complex features formerly reserved for biology. Such systems possess a complicated structure interacting internally via its parts and with its environment. These systems are evolved or historical in the sense that their organization has not been primarily designed, but has emerged gradually through accreted solutions to particular problems. They are thus extremely difficult to re-engineer given that what may now be undesirable elements support essential ones.[23]

In some sense these historically extremely recent human-made systems can even be said to possess the types of agency found in natural systems such as forests or swarms. At the moment it is still the decisions of individual human beings who sit at the root of these systems, but we are gradually being accompanied by artificial agents, and in some domains, such as financial markets, humans are being supplanted by AI.[24]

Perhaps most importantly, complex systems, both of the biological and the human created sort appear to be far too dynamic and reliant on networks of internal and external feedback to be manageable via centralized control. Instead it is the interaction of the parts themselves in coordination and competition that gives rise to a complex system's order and behavior.

The recognition that the technological transformation brought about by science was creating a type of society where top-down control was no longer tenable is by no means new. Here we see the root of 20th century economists such as Fredrick Hayek who argued for the wisdom of "the market" as a source of solutions to social problems over deliberate interventions by the state.[25] This move from deliberate action in the face of problems "from above" to one favoring solutions emerging from the actions of individuals, "from below" was not one that emanated from 20th century conservatives such as Hayek alone. Instead it has had articulate advocates from progressives, such as Jane Jacobs,[26] who have seen in such emergent solutions not just a more humane way of organizing society than its alternative, but a way of finding better answers to problems than anything likely in solutions derived from the top down. The process of rising complexity, however, has perhaps now far outgrown these earlier prescriptions.

It might be the case however those who in the 20th century were drawn to such bottom up systems were blinded by a vision of society that is ceasing to exist, the small market of independent merchants, or the neighborhood. This was a society where the decisions of individuals remained the key source of the collective knowledge about the society itself and remained the most reliable source for decision making. The multitude of globe straddling interacting systems found in our own time are much different.

They are examples of what the philosopher Timothy Morton has called "hyper-objects", systems that we cannot hope to fully understand or control even though we are completely enmeshed and dependent upon them.[27]

In the emergence alongside natural hyperobjects of human made systems that rival their scale Morton thinks we can witness the birth of the Anthropocene. When human civilization has reached such a capacity to reshape the earth that it now resembles a geological force. Our energy system would be one such hyperobject, as would our food system, along with internet. Such systems do indeed embody a kind of collective intelligence, a mind, even when that system is leading to self-destructive outcomes, such as climate change. Besides such destructive environmental outcomes the negative impact of such hyperobjects is to both severely constrain individual freedom, at least when it comes to the system itself, and, above all, make universal understanding (logos) an increasingly impossible goal.

If the scientific and democratic revolutions can be said to share a mutual world-view it is that with increasing knowledge would come increasing freedom. Scientists would by coming to know the world be better able to affect it, just as citizens with a better knowledge of the world would be able to choose the best course for their societies.

What, since at least the industrial revolution, we've experienced is a world where an increase in individual freedom, the ability of an individual to pursue self-selected ends, has been bought in exchange for loss of control over the systems via, and in light of which, the individual pursues her goals. Systems have become distant and abstract, and control over them, to the extent that it exists and doesn't merely emerge from interacting individuals, has become highly concentrated, asymmetric and sharply delineated.[28]

We already legally treat such entities, the corporations which serve as the core of the modern economy as *persons,* and they often far outlive the distributed network of individuals who now direct their operations. At some point, however, it might become possible to automate this human direction entirely so that AI becomes in the words of science writer Alexis Madrigal "the avatar of corporate persons".[29]

At the same time AI is becoming a tool of human management many corporations seem bent on fully injecting intricately calibrated levels of control into whatever area comprises their domain. Communications and retail companies pursue the creation of and deployment of intelligent devices that can intimately monitor, respond and influence customers,[30] while biologically centered firms seek to engineer life to make it conform to human wishes.[31] Having achieved the extraction of the idea of will from the rest of nature we are on the verge of restoring it as nature itself becomes infused with human will or reified into the form of AI that, at least for now, is a reflection of human will. Such agency will be encountered everywhere from the genetically engineered life that will surround us, the atmosphere we will have shaped through dereliction or design, and the intelligence that will be woven into even the smallest corners of the built environment.[32] It will even be confronted in those spaces of wilderness we now consider an escape from civilization.[33]

Our experience of an earth suffused with agency may even accelerate the reintro-duction of the idea of will into the universe itself, and especially unto those questions in which traditional science has, so far, been unable to answer. We might be con-fronted with an increasing number of situations in which the line between some nat-ural phenomenon we cannot explain and the possibility of that phenomenon being an artifact of some alien intelligence, perhaps artificial and far more advanced than our own, cannot be definitively drawn.[34] This projection of agency upon phenomenon whose origin we are unable to explain using current scientific theory might even lead cosmology itself down a rabbit hole in which even the laws of physics themselves are explained in reference to some form of mind and agency beyond human ken.[35]

Thus the world might become infused with agency, though in the sense that the individual is surrounded by it rather than experiencing it as a world shapeable by her unique will. There are steps we can take to avoid the worst aspects of such a fate, especially when it comes to the question of artificial intelligence, which is likely to increasingly be tasked with controlling the systems upon which we depend yet are unable to control.

If society continues to fail to educate citizens in how these artificially intelligent systems work, or fails to empower them with the ability to influence their program-ming, or even just inform the non-programming public that there is a human interest behind every bot, the world might again be perceived as being under the suzerainty of capricious and cruel gods or perhaps be understood as the mere tool of some imag-ined human conspiracy.[36] Freedom as we currently understand it will only survive in such a world if citizens are granted enhanced abilities to steer, shape, and even redesign the complex systems upon which we have come to depend.

What makes such a preservation of freedom even more daunting is that the erosion of this freedom, when it comes to systems, proceeds at the same time the individual is losing the capacity for logos. As long as we remain human we will possess mind or nous, but logos, the ability to rationally and comprehensively understand the kosmos is much more vulnerable. A greater commitment to wide-ranging science education might address some of this, but not all, for in a sense, the problem of incomprehensibility emerges from the success of science as a social endeavor itself.

The more we learn of the world the less a comprehensive view of it seems possible given the limitations of the individual human mind.[37] We should hope that our seem-ing possession of logos may has not been a unique historical interregnum that began when the pre-Socrates set out to explain the world using rational theories graspable by the individual. For should we end up with a world infused with agency, much of it hidden and incomprehensible to even the most educated individuals, where much of what happens can only be explained by the conflict between non-human rival forces (the state, corporations, AI), or some form of intelligence beyond our capacity to understand, then we will have in some sense come full circle. It would be a less enlightened and less free world, though perhaps not without its wonders, a world that would be instantly recognizable to my imagined Aristagoras.

End notes

1. "The so-called "honeybee conjecture. Proposed by Marcus Terentius Varro in 36 BC (thus centuries later than my imagined dialogue) it was only prove correct in 1999 by Thomas C. Hales. Hales, T. C. "The Honeycomb Conjecture." Discrete & Computational Geometry 25, no. 1 (2001): 1–22.

2. The views of pre-Socratics presented in the dialogue adapted from: McCoy, Joe. *Early Greek Philosophy*. Washington: Catholic University of America Press, 2013.

3. Scranton, Roy. *Learning to die in the Anthropocene: reflections on the end of a civilization*. San Francisco, CA: City Lights Books, 2015, pp. 113–114.

4. Dolnick, Edward. *The clockwork universe: Isaac Newton, the Royal Society, and the birth of the modern world*. New York, NY: Harper, 2011.

5. Gould, Stephen Jay. *Wonderful life: the Burgess Shale and the nature of history*. New York: Norton, 2007. Since Gould's death Henry Gee has taken up the mantel of arguing for the non-deterministic, non-progressive nature of evolution. See: Gee, Henry. *The accidental species: misunderstandings of human evolution*. Chicago, Ill: The University of Chicago Press, 2015.

6. England, Jeremy L. "Statistical physics of self-replication." *The Journal of Chemical Physics* 139, no. 12 (2013).

7. Gray, M. W., J. Lukes, J. M. Archibald, P. J. Keeling, and W. F. Doolittle. "Irremediable Complexity?" *Science* 330, no. 6006 (2010): 920–21.

8. Arbesman, Samuel. *Overcomplicated: technology at the limits of comprehension*. New York, NY: Penguin Randon House, 2017.

9. Tononi, Giulio. *Phi: a voyage from the brain to the soul*. New York: Pantheon, 2012.

10. Montgomery, Sy. *The soul of an octopus: a surprising exploration of one of the world's most intriguing characters*, London: Simon & Schuster, 2016.

11. Moroz, Leonid L., and Andrea B. Kohn. "Independent origins of neurons and synapses: insights from ctenophores." *Philosophical Transactions of the Royal Society B: Biological Sciences* 371, no. 1685 (2015): 20150041. Accessed September 2, 2017. https://doi.org/10.1098/rstb.2015.0041.

12. Giovannetti, Manuela, Luciano Avio, Paola Fortuna, Elisa Pellegrino, Cristiana Sbrana, and Patrizia Strani. "At the Root of the Wood Wide Web." *Plant Signaling & Behavior* 1, no. 1 (2006): 1–5. Accessed September 2, 2017. https://doi.org/10.4161/psb.1.1.2277.

13. Wohlleben, Peter, and Tim F. Flannery. *The hidden life of trees: what they feel, how they communicate: discoveries from a secret world*. Berkley, CA: Greystone Books, 2016.

14. Lanier, Jaron. *You are not a gadget a manifesto*. New York: Knopf, 2011, p. 154.

15. Russell, Stuart J., and Peter Norvig. *Artificial intelligence a modern approach*. Boston: Pearson, 2016.

16. Wolfram, Stephen. *A new kind of science*. Champaign, IL: Wolfram Media, 2002.

17. Dennett, Daniel Clement. *Freedom evolves*. London: Penguin, 2007.

18. Ismael, Jenann. "Causation, free will, and naturalism." *Scientific metaphysics* (2013): 208–235.

19. Quoted in Gleick, James. *Time travel: a history*. New York: Pantheon Books, 2016, p. 260 Here Gleick makes an impassioned case against determinism.

20. Arendt, Hannah, and Hannah Arendt. "What is Freedom?" in *Between past and future, six exercises in political thought*, 143–71. New York: Viking Press, 1961.

21. Eisenstein, Elizabeth L. "The book of nature transformed" in *The printing revolution in early modern Europe*. Cambridge: Cambridge University Press, 2013.

22. Blom, Philipp. *Enlightening the world: Encyclopédie, the book that changed the course of history*. New York: Palgrave Macmillan, 2005.

23. Ibid. Arbesman

24. Tett, Gillian. "Welcome to a wild world of robot investing." Financial Times. August 27, 2015. Accessed September 2, 2017. https://www.ft.com/content/0538655a-840b-11e5-8e80-1574112844fd.

25. Hayek, F.A. "The uses of knowledge in society." *The American Economic Review* 34, no. 4 (September 1945): 519–30. Accessed September 2, 2017. http://home.uchicago.edu/~vlima/courses/econ200/spring01/hayek.pdf.

26. Jacobs, Jane. *The death and life of great American cities*. New York: Modern Library, 2011.

27. Morton, Timothy. *Hyperobjects: philosophy and ecology after the end of the world*. Minneapolis: University of Minnesota Press, 2014.

28. "A giant problem." The Economist. September 17, 2016. Accessed September 02, 2017. https://www.economist.com/news/leaders/21707210-rise-corporate-colossus-threatens-both-competition-and-legitimacy-business.

29. "Episode 3: Ask not what the robots can do for you." Radio Atlantic. August 4, 2017. Accessed September 02, 2017. https://www.theatlantic.com/podcasts/radio-atlantic/.

30. Sterling, Bruce. *The Epic Struggle Over the Internet of Things*, e-book, Strelka Institute for Media, Architecture and Design, 2014.

31. Anthes, Emily. *Frankenstein's Cat: Cuddling Up to Biotech's Brave New Beasts*. London, UK: Onworld Books, 2013.

32. The core idea behind the concept of the Anthropocene is of a world shaped by human will. See: Ackerman, Diane. *The human age: the world shaped by us*. Toronto, Ontario, Canada: HarperCollins Publishers Ltd, 2015. The idea that the IoT (Internet of things) where everyday objects are embedded with their own intentions along with the agendas of the companies that make them might result in a resurgence of animistic dread was aptly pointed out by Marcelo Rinesi. "The price of the Internet of Things will be a vague dread of a malicious world." Institute for Ethics and Emerging Technologies. September 25, 2015. Accessed February 26, 2017. http://ieet.org/index.php/IEET/more/rinesi20150925.

33. Duane, Daniel. "Opinion | The Unnatural Kingdom." The New York Times. March 11, 2016. Accessed September 02, 2017. https://www.nytimes.com/2016/03/13/sunday-review/the-unnatural-kingdom.html?mcubz=3.

34. Wright, Kimberly CartierJason T. "Have Aliens Built Huge Structures around Boyajian's Star?" Scientific American. May 1, 2017. Accessed Septem-

ber 02, 2017. https://www.scientificamerican.com/article/have-aliens-built-huge-structures-around-boyajian-rsquo-s-star/.

35. Scharf, Caleb. "Is Physical Law an Alien Intelligence? - Issue 42: Fakes." Nautilus. November 17, 2016. Accessed February 26, 2017. http://nautil.us/issue/42/fakes/is-physical-law-an-alien-intelligence.

36. Pasquale, Frank. *Black box society: the secret algorithms that control money and information*. Cambridge, MA: Harvard University Press, 2016.

37. Lem, Stanisław, and Dariusz Fedor. *Summa technologiae*. Warszawa: Agora, 2010.

Chapter 15
No Ghost in the Machine

Alan M. Kadin

The prevalent pre-scientific paradigm for understanding nature focused on design or intention, even for inanimate objects. This approach was debunked by Newton for physics, and by Darwin for biology. But belief in the unique supernatural nature of human intelligence is still widespread. I argue that biological intelligence is due to simple evolved structures based on neural networks, without the need for any new physical mechanisms (quantum or classical) or a "ghost in the machine". Humans see agency and intent everywhere, because we are programmed to do so. The conscious mind may turn out to be a virtual reality simulation that is largely illusory. Furthermore, these structures may be emulated in artificial neural networks, to create true artificial intelligence.

> Everything should be made as simple as possible, but not simpler.
>
> Attributed to Albert Einstein [4]

15.1 Introduction

This year's essay contest for the Foundational Questions Institute on "Wandering Towards a Goal" contrasts "mindless mathematical laws" with "aims and intentions", and asks how this apparent paradox can be resolved. This is just a restatement in modern scientific language of the classic philosophical question about the origin of life and human intelligence. Or in religious terms, is the human soul special and distinct from the inanimate physical universe? Does this question really belong in a physics-based essay contest?

I argue below that if one properly understands the scientific basis for physics, biology, psychology, and computer science, there is no paradox. But the superficial

A. M. Kadin (✉)
Princeton Junction, NJ, USA
e-mail: amkadin@alumni.princeton.edu

© Springer International Publishing AG, part of Springer Nature 2018
A. Aguirre et al. (eds.), *Wandering Towards a Goal*, The Frontiers Collection,
https://doi.org/10.1007/978-3-319-75726-1_15

philosophical implications of each of these may be misleading. For example, the mathematical equations of simple physical systems lead to causality and determinism, but real physical systems have noise, and complex physical systems tend to be chaotic and unpredictable. The paradigm of causal control in complex systems is an illusion, and reality is more subtle.

Similarly, biological evolution is predicated on random mutations, but leads to complex structures and organisms that appear to be "intelligently designed". So biological design, too, is an illusion, which is explainable in terms of blind adaptation to complex environments. This biological tendency toward complexity may seem incompatible with the thermodynamic tendency toward increasing entropy in a closed physical system near equilibrium, but biological systems are neither closed nor in equilibrium.

Regarding human intelligence, people feel that they are independent rational agents, but in reality, most of human behavior is subconscious and irrational. In classical computer science, a computer is an arithmetic engine with external memory, but most modern computational problems related to "artificial intelligence" are more akin to matching patterns than to arithmetic calculations, and the same is true for natural intelligence.

Further, I argue that consciousness itself reflects an evolved brain structure that is not uniquely human, and provides an adaptive system capable of making rapid decisions based on simplified models and incomplete data. An analogy may be with a dynamic virtual reality environment that integrates and synthesizes data from diverse sensor inputs and generates motor outputs that are reflected within the same virtual environment. I further suggest that a similar consciousness engine may be emulated artificially.

The title of this essay comes from the 1967 book by Arthur Koestler, "The Ghost in the Machine" [14], which in turn came from a phrase used by philosopher Gilbert Ryle to describe mind-body dualism. My view is that the subjective experience of consciousness reflects the brain activity associated not with the entire brain, but only that small portion that is projected into this self-conscious virtual reality construct. In the following sections, the prevailing scientific and philosophical paradigms are reviewed, and the illusions and paradoxes resolved. The guiding principles throughout will be simplicity and unity, as informed by Occam's Razor, as well as Einstein's dictum above [4].

15.2   Physics, Mathematical Determinism, and the Illusion of Control

In pre-modern physics, as exemplified by the work of Aristotle, the behavior of an object, animate or inanimate, was believed to be driven by its natural tendencies, in a qualitative rather than a mathematical sense. This general doctrine is known as teleology, from the Greek for end, i.e., goal. Fire tends to rise, while massive objects fall because that is their nature. This approach was a partial attempt to get away from the anthropomorphic tendency of pre-scientific peoples that all objects in nature are driven by spirits.

A further general principle in pre-modern physics was that humans are the center of the universe, with the earth and the heavens having fundamentally different laws. For example, objects on earth tend to move in straight lines, while objects in the heavens tend to move in perfect circles. But in order to get this geocentric model to work, Ptolemy modeled the motion of planets using epicycles, i.e., circles within circles. When this was done properly, it worked quite well, but it was both complicated and arbitrary. In fact, the Ptolemaic epicycles worked better than the Copernican model of circular orbits around the sun (because the actual orbits are elliptical). As Steven Weinberg has pointed out in his recent book [24], the Copernican model was accepted not because it was more accurate, but rather because it was simpler and more unified.

Isaac Newton took simplification and unification much further, by developing a quantitative theory of universal gravitation on both the earth and the heavens, and by developing the mathematics (differential equations) needed to do the calculations. The conventional paradigm for classical Newtonian physics is the "Clockwork Universe", where everything in the universe follows mathematical trajectories that are fully deterministic and completely predictable. This was very influential in the sciences and beyond, including with the doctrine of philosophical determinism, that the future is completely defined and can never be altered.

It is widely believed that this clockwork paradigm held until quantum mechanics introduced indeterminacy in the early 20th century, but this is incorrect. The entire fields of statistical mechanics and thermodynamics were developed in the 19th century to deal with uncontrolled thermal noise in classical many-body systems, which are microscopically deterministic but macroscopically random. This led to the concept of entropy, which is a measure of the probability of a given physical many-body configuration. For a closed system in equilibrium, entropy must always increase, leading to more probable configurations. While the microscopic equations of classical physics are time-reversible, irreversibility associated with increasing entropy enables one to define the arrow of time. Adaptive biological systems may lead to increased complexity, but this is not in conflict with increasing entropy overall (see below).

More recently in the 20th century, classical nonlinear dynamical systems were shown to be highly sensitive to initial conditions in a way that is practically unpredictable (the "butterfly effect"). The mathematical theory of deterministic chaos was developed to describe the dynamics of some of these classical nonlinear systems. Furthermore, macroscopic systems often include complex feedback loops that largely decouple the macroscopic behavior from microscopic degrees of freedom. While textbooks emphasize the mathematical certainty of simple dynamical systems, the behavior of real classical systems is more subtle and much less predictable.

This is illustrated in the block diagrams shown in Fig. 15.1. Figure 15.1L shows a classical causal sequence of A causing B causing C. In contrast, Fig. 15.1R shows a more complex set of interactions: each component affects the others, but also is affected by them via back-action. Furthermore, each component also acts on itself

Fig. 15.1 Comparison of simple causal sequence to more complex set of interacting systems. (Left) Causal sequence of A causing B causing C. (Right) Interacting systems A, B, C, and N, where N represents uncontrolled factors such as noise

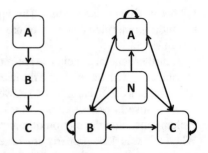

(recursion), and is affected by noise sources N. For example, consider Fig. 15.1R in the context of global warming. The atmosphere might be system A, the biosphere might be system B, and the carbon reservoir in fossils fuels might be system C. The noise N represents those factors, local and global, which are not controlled. The interactions between these systems can provide the basis for a complex model, which is still oversimplified—there is also radiation input/output and the influence of the oceans. While the near-term trends are clear, the accumulation of uncertainties and model inadequacies make long-term predictions less reliable.

Philosophical determinism is frequently applied to human interactions, where there are not even good mathematical models. It makes little sense to derive human-level determinism from the mathematical trajectories of planets or atoms. The philosophical doctrine of free will is often contrasted with determinism, on the level of individual humans. Where determinism suggests that the future is completely predictable, free will suggests that one's future is undetermined until one makes a decision. But both determinism and free will are really straw men; as individuals, we are under varying degrees of external and internal constraints. You are free to choose your breakfast, but the climate of the earth is beyond your control. Furthermore, we tend to see ourselves as free agents, but we actually have much less control than we think we do. Control is just another illusion. The issue of agency will be addressed further in Sect. 15.4.

Quantum mechanics incorporates indeterminacy even on the microscopic level, which is different from classical uncertainty. However, the major effect of quantum uncertainty on the macroscopic level is simply to introduce another source of noise (quantum noise), in addition to thermal noise. There are other more exotic effects of quantum uncertainty, such as quantum entanglement, but it is unclear whether these have any significant impact on macroscopic systems at ambient temperatures. A further exotic aspect is quantum measurement theory, which focuses on the role of the observer in changing the quantum state. It has been suggested that these poorly understood paradoxical aspects may provide a basis for aspects of the human mind and consciousness, but this would seem to be yet another example of anthropocentrism. In contrast, I have proposed [8, 10] an alternative picture of quantum mechanics that avoids the paradoxes of entanglement and measurement. This is addressed briefly in the End Notes.

15.3 Biology, Evolution, and the Illusion of Design

Living organisms have long been regarded as different from the rest of nature. Indeed, it was widely believed that biology did not follow the same physical laws; this is the basis for the discredited doctrine of Vitalism. In fact, biological systems appear to be intricately designed machines, without a self-evident set of instructions. It was not until the middle of the 20th century that the DNA genetic code was discovered, providing a digital program of instructions for growing an animal, plant, or microorganism.

But the greatest breakthrough in biology occurred in the 19th century with Darwin's Theory of Natural Selection, commonly known as evolution. Evolution is both the simplest and the most profound concept in all of science; all it requires are exponential reproduction and genetic noise, interacting with an environment. But prior to Darwin, this was hidden for centuries, because of the illusion of intelligent design.

As shown in Fig. 15.2, a guided design (Fig. 15.2T) follows a specific plan towards a goal, whereas in natural selection (Fig. 15.2B), all possible changes to the current design are generated automatically, but only those that provide improved adaptation to the environment and reproductive success are preserved for future generations. These diagrams appear somewhat similar, but have critical differences. In particular, only a guided design can produce a radical redesign in a single step. In contrast, the unguided design of natural selection can only make minor modifications per generation, each of which must be adaptive. But evolution can often be quite fast, since it is both massively parallel (all organisms) and serial (once per generation). Although biological systems may initially appear to be well designed, they are locally rather than globally optimized. Their design limitations follow from their earlier history. This point was emphasized by Gould [5] in "The Panda's Thumb". The thumb of the panda is not a thumb at all, but rather a modified wrist bone. The thumb bones had been eliminated in the earlier evolution of the panda's ancestors.

Random mutations produce the variation needed to generate environmental adaptation. In addition, random "genetic drift" in small populations may sometimes play a role in evolution. But in most cases, evolution is deterministic and can be modeled mathematically, although real natural environments tend to be much more complex and dynamic than simple models. There is no underlying goal or intent, apart from survival. The primary feature that distinguishes biological systems from physical systems is exponential reproduction, based on the digital code of DNA. All life on earth has the same genetic code, and the initial origin of DNA-based life is one of the remaining questions in the theory of evolution. It has been suggested that an RNA World based on the catalytic properties of RNA might provide a precursor to the genetic code, thus enabling the origin of life on earth.

Evolution is driven by random variation and guided only by the environment for a given organism, where the environment includes a multiplicity of other organisms. There has been a move toward increasing complexity in evolution, but this is not complexity for its own sake. Indeed, excessive complexity is not favored by natural selection; like theories of physics, following Einstein's dictum, life should be as

Fig. 15.2 Comparison of guided and unguided design pathways. (Top) Guided design pathway, whereby a product is designed for a functionality or market. (Bottom) Unguided design pathway, as in biological evolution of an organism

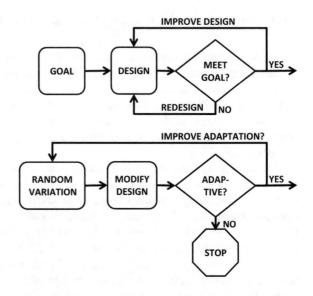

simple as possible, but no simpler. Complex organisms developed only because they are better adapted to compete successfully in complex environments.

If animal bodies have evolved complexity to compete more successfully, the same should be true for animal brains. We are accustomed to comparing the sizes of brains, but brain organization is arguably more critical. Brain structures evolved because they improved environmental fitness and survival of progeny. These include not only those structures associated with sensor input or muscular coordination, but also those associated with intelligence and consciousness. Brain structures involved with intelligence (adaptive learning in complex environments) may be quite widespread in the animal kingdom, particularly among vertebrates. Peter Godfrey-Smith recently argued [6] that cephalopods (octopi and squids), mollusks unrelated to vertebrates, may also have complex brains capable of adaptive learning. This may provide an example of convergent evolution. Finally, while human consciousness involves language and abstraction, it may not be qualitatively different from that in "lower" organisms.

15.4 Psychology, Cognition, and the Illusion of Agency

We all feel that we can understand the human mind based on our own thinking. We identify ourselves as individual agents who perceive the world around us and are in control of our own actions. Indeed, Rene Descartes took this as the basis for his philosophy: "Cogito, ergo sum", or in English, "I think, therefore I am." But this makes the mind a distinct object separate from the body and from the physical world, i.e., mind-body dualism. This outlook is embedded in our sense of self and

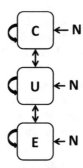

Fig. 15.3 Block diagram of control in corporation or human mind. In the corporate example, C may represents the Chairman of the Board, U the underlings, and E everyone else. In a model of the mind, C may represent the conscious mind, U the unconscious mind, and E the environment. (N represents random noise.) In both cases, top-down control and agency may be illusory

in our language for describing it, but there are good reasons to believe that this is mostly an illusion. Research in psychology and cognitive science has shown that our self-perceptions can be quite deceptive.

Historically, one response to this dualism was to go to the other extreme and deny the existence of an internal mind. This is the basis for behaviorism, of which B.F. Skinner was a leading 20th century proponent. This deals with an animal or even a human as a "black box" with inputs and responses. More recent approaches in cognitive science have recognized that there are internal cognitive structures (both conscious and unconscious), and have identified fallacies of thinking that reflect aspects of these internal structures.

This was explored in the psychological research of Daniel Kahneman, as summarized in "Thinking, Fast and Slow" [13]. This showed by careful experiments that people are not really the rational free agents they think they are. This has been particularly influential in providing the basis for behavior economics. Kahneman identified two distinct systems at work in the human mind: System 1 and System 2. System 1 is the unconscious mind that does things automatically without us having to think about them. System 2 is the conscious mind, which requires deliberate attention and thought. Most of our actions are actually done by System 1, even though we firmly believe that they were rational decisions made by our conscious mind. System 2 operates with a simplified model and a coherent narrative, and when the model appears inconsistent ("cognitive dissonance"), the perceptions may be altered to maintain a consistent picture.

Kahneman uses the analogy with a Chairman of the Board of a corporation, who thinks that he runs the entire operation. Most of the work is actually done by his underlings, generally without even asking the chairman. The underlings give the chairman credit for successful operation, building up his ego, and he believes it. In this context, the chairman represents the conscious mind, which has an inflated sense of its own importance, and is largely unaware of the existence of the unconscious mind. This is summarized in the block diagram in Fig. 15.3.

It seems that the human mind is preprogrammed to identify agency, both in ourselves and in others (and even in inanimate objects!). These agents are central to a simplified model of the world, which filters all our perceptions. Further, events do not just happen in isolation, but represent a causal sequence, a coherent story in progress, with heroes, villains, and value judgements. The conscious mind operates on this simplified model, although we are conditioned to believe that this is the real world. Simple goals are identified which promote basic urges such as survival, pursuit of pleasure, and avoidance of pain. The brain organization that supports this evolved to enable rapid decisions in complex dynamic environments, with incomplete information. In a fight-or-flight situation, indecisiveness is not adaptive. I am not suggesting that we are living in "The Matrix", but perhaps we are each living in our own Matrix, with only a partial mapping to those of other people and to the real world.

15.5 Computation, Neural Nets, and the Illusion of Intelligence

Ever since digital computers were first developed in the 1950s, they were commonly thought of as "electronic brains". But traditional computers are actually quite different from brains, both in structure and capabilities. However, recent research in alternative computer architectures, combined with research on the brain, has shown that "neuromorphic" computer architectures may finally be emulating brains more closely (see also [19]).

This difference can be seen in Fig. 15.4, which compares a traditional von Neumann computer architecture to a neural network. The von Neumann architecture in Fig. 15.4T extended a simple mathematical model of a computer proposed by Alan Turing, and was brought to practical fruition under the direction of John von Neumann at the Institute for Advanced Study in Princeton (see [3]). This architecture consists basically of an arithmetic engine with a control program and memory. Virtually all practical computers since then have had the same basic structure. Even modern supercomputers with multiple processors and memory hierarchies are based on this original design.

Compare this to a basic design of an artificial neural network (or neural net) shown in Fig. 15.4B. This represents an array of artificial neurons (which may be electronic elements) organized into layers and connected with "synapses". The strength of a given synapse may be changed according to a procedure of iterative training or learning, rather than an imposed program. The collection of synapse strengths correspond to memory, but are distributed throughout the system, rather than localized in a single module. Furthermore, there is no central processing unit; the processing is also distributed. This structure is similar in several respects to the interconnections between biological neurons, although it is much simpler.

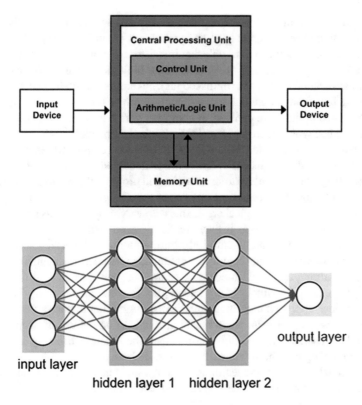

Fig. 15.4 Computer architectures for (Top) classic von Neumann computer and (Bottom) artificial neural network (from Wikipedia)

The von Neumann architecture represents a universal computer, so that it can be used to simulate any other computer architecture, including a neural net. But such a simulation may be inefficient and slow. In fact, a traditional processor works very well in doing mathematical calculations, whereas brains do mathematics rather poorly. On the other hand, brains have evolved to match patterns, and do this very well. Pattern matching is not limited to image recognition, but occurs in all sensory processing, memory retrieval, language translation, and a host of correlation and optimization problems. Recent research has shown that neural nets with many "hidden layers" between the input and output can be trained to be particularly efficient in learning to match patterns; this is known as "deep learning". More hidden layers enable more abstract correlations among inputs, which in turn enables more flexible recognition of a wider variety of complex patterns. The learning process itself is adaptive in a way that is similar to evolution. The environment selects those variations that are most effective in matching the training patterns.

Another important aspect of brains is that they are composed of basic elements (the neurons) that are slow, noisy, and unreliable. But despite this, brains have evolved to respond quickly in complex environments, by taking advantage of distributed computing with massive parallelism, and neural nets are uniquely capable to taking full advantage of this. It is remarkable that transistors, the basic elements of modern computers, are now a million times faster than biological neurons (characteristic times of nanoseconds rather than milliseconds), but brains are still far faster and more energy-efficient than traditional supercomputers in the types of pattern matching tasks for which brains have evolved. A computer with a brain-like organization but electronic speed would represent a major technological breakthrough.

The field of "artificial intelligence" is almost as old as computers themselves, but it has long fallen far short of its goals. The traditional method of artificial intelligence is to devise a list of rules about a particular topic, and program them into a conventional computer. But knowledge of a fixed set of rigid rules is not what we generally mean by intelligence. Indeed, biological organisms with very simple nervous systems, such as worms or insects, behave as if they are "hard-wired" with rule-driven behavior, and are not regarded as intelligent at all. Furthermore, sophisticated responses to dynamic environments cannot be fully programmed in advance; there must be a strong element of learning involved. In contrast, neural nets can learn how to recognize a cat [18], for example, without being told explicitly what defines a cat. The newer "deep learning" approach to artificial intelligence is starting to have a major impact on technology (see, for example, a recent article [16] on the use of deep learning for Google Translate). In traditional computers, the greatest difficulty is found in writing and debugging the software. In contrast, in both natural intelligence and the newer approaches to machine learning, the software is generated automatically via learning. No programmer is necessary; as with evolution, this is unguided.

15.6 Minds, Dreams, and the Illusion of Consciousness

The most persistent illusion associated with human consciousness is that there must be an immaterial spirit. But this is clearly a remnant of pre-modern religious thinking, where everything is driven by immaterial spirits. Efforts to assert that somehow consciousness emerges from brains of a certain scale or complexity (see e.g., Teilhard de Chardin) are misdirected. Biological vision did not arise simply from scale or complexity; it required evolution of structures with specific functionalities. Why should we expect consciousness to be different? Several philosophers, such as Daniel Dennett [1], have been arguing for some time that consciousness is an illusion, but they have been less clear as to the actual structure behind the illusion. There have also been various proposals, mostly not very specific, for machine consciousness [17].

Consider a preliminary outline of the structure of a consciousness organ, which might be based on neural networks. While many aspects responsible for consciousness are hidden from view (either external or internal), some key functional requirements should be clear. Consciousness involves a self-identified agent following a continuous, causal narrative in a dynamic environment. The environment must inte-

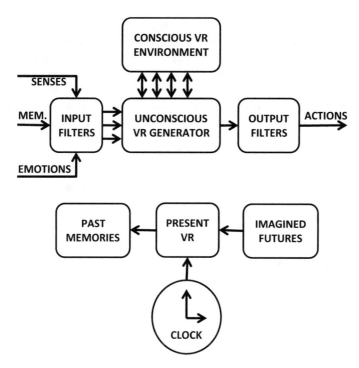

Fig. 15.5 Suggested organization of autonomous natural or artificial intelligence. (Top) Unconscious mind integrates pre-filtered senses, emotions, and memories, and generates VR environment with active agent. (Bottom) Present VR frame is time-stamped and shifted into the past, and a future VR is selected to overlay the present

grate various sensory modalities with relevant emotions and memories. This is shown in Fig. 15.5T, which shows consciousness as a "virtual reality" (VR) construct created from filtered input data, and representing a simplified dynamic model of the reality presented to an individual. In addition, this must also be linked to a short-term dynamic memory module, containing the recent past, and a predictive model of one or more near futures, as shown in Fig. 15.5B. A clock time-stamps the present frame and shifts it to the past, while selecting a possible future. This ensures that perceived time is a central element in consciousness.

For example, a conscious visual representation of a rose is not just a portion of a larger two-dimensional image. Rather, it is an object embedded in a three-dimensional space, which represents the concept of a rose and links to abstract attributes and memories of images and aromas. This may be analogous to a PDF file of a scanned document which is subjected to optical character recognition (OCR). Such a file contains an image of a given page, but the image of each character is also associated with the functional representation of the character. Now imagine that the document also contains embedded links to definitions of words and other relevant documents, and one starts to have the richness necessary for consciousness. This would require that the consciousness organ represent the central nexus of a network linking the filtered inputs and the control outputs. Indeed, recent research [15] has found direct evidence for a network of this type in the cerebral cortex, which may represent such a conscious nexus. In contrast to a static environment of a document, the conscious environment is highly dynamic, and is rapidly updated in time. We have the sense that time is continuous, but the actually memory elements are probably discrete. That is the case with virtually all technological video representations, which are perceived as continuous if they are fast enough.

But the most critical aspect to explain is the subjective sense of agency. I would suggest that this is due to an adaptive neural net, primed from infancy to recognize self-agency and causality, and also to recognize external agents. The dynamic environmental model is built around such actors. As Shakespeare said [21], "*All the world's a stage, and all the men and women merely players...*" The sense of self is merely the internal sensation of activation of brain circuits in the consciousness nexus. Note that recognition of agency is really a form of temporal pattern recognition. We live in a world that is continuous and causal, so that construction of simplified causal models based on observed temporal correlations may often be highly adaptive. Furthermore, this is a case of dynamic learning; the mind learns to generate and refine a simplified model which maintains effective interaction with the environment.

Perhaps the strongest evidence for a dynamic VR generator occurs during dreams, when input and output are isolated from the real world. Dreams include self-agency and coherent narratives in complex environments. Allan Hobson [7] has recently suggested that dreams may represent adaptive training and optimization of the consciousness engine, whereby memories are consolidated or forgotten, as appropriate. This may be in accord with other recent research that has shown that brain interconnections grow dramatically during daytime activity but are selectively pruned back at night [23].

An important aspect of investigating consciousness is how we can identify it in either artificial or biological systems. Consciousness may be an evolved adaptive structure that is widespread among animals, even without language and other abstractions. Such a conscious structure may never have been present in any artificial intelligence system to date, but it could be constructed in the near future. How can we tell? Alan Turing proposed the "Turing Test" [22] for artificial intelligence based on asking a series of questions. However, given the human tendency to see agency, this may be deceptive. Many years ago, Joseph Weizenbaum of MIT developed the ELIZA system to simulate natural human conversation without any underlying understanding. So a non-verbal method of exploring consciousness might be preferable. Science fiction writer Philip K. Dick wrote "Do Androids Dream of Electric Sheep?" [2], which later became the basis for the film "Blade Runner." Perhaps we will know that we have made conscious machines when we can observe them dream.

15.7   Conclusions

The stated question in this essay contest, *"How can mindless mathematical laws give rise to aims and intention?,"* implicitly assumes that human behavior should be ultimately derivable from particle physics. This is entirely wrong, on several levels. The problem is that there are several illusions implicit in our thinking that impede our understanding, illusions of control, design, agency, intelligence, and consciousness. The conscious mind is an evolved, adaptive brain structure that hides more than it reveals.

A second point is that the paradigm of natural selection is central not only to biology, but to psychology as well. Neural networks are capable of learning and adaptation to complex environments, and the conscious mind represents a simplified dynamic model of the environment. Goals and intentions are abstract representations of adaptive programs that can promote individual well-being and success. Human behavior can be predictable and causal, subject to statistical variation. There is no ghost in the machine, and no paradox.

Third, true artificial intelligence and consciousness can be emulated in a properly designed electronic system, probably within about 20 years. Such an autonomous system can distinguish self, other agents, and objects, and can create a simple causal narrative of changes in its environment. Further, it is likely that most of our food animals (except perhaps shellfish) may also be conscious. Together, these will profoundly impact our sense of what it means to be human. We evolved in small family bands of hunter-gatherers; our conscious minds may be too simple to deal with a complex modern world. Some sort of augmented artificial intelligence may be required for humans to survive to the next century. Let us hope that this can incorporate more of the positive human attributes of rationality and cooperation, rather than the negative human attributes of war and xenophobia. Notwithstanding micro-determinism, the future of humanity is still undetermined. Only time will tell.

End Notes

I have argued here that direction and intelligence in nature follow from adaptation in a complex dynamic world based on the foundations of classical physics, without the need for anything either quantum or supernatural. Elsewhere, I have argued that the foundations of quantum mechanics have been profoundly misunderstood, and that quantum mechanics properly provides a unified foundation for all of physics. The comments below describe some implications of this for the nature of time and for the future of computing.

A. Quantum Waves and the Nature of Time

Time is central to physics, biology, psychology, and computation. But is it the relativistic spacetime of Einstein, or the subjective time of our internal chronometers?

I have suggested [12], and in an earlier FQXi essay [9], that time is defined on the microscopic level by quantum waves, and everything else follows from that.

In the Newtonian clockwork universe, time is abstract and universal. In Einstein's universe, time becomes part of abstract spacetime, which is non-universal and inhomogeneous. Einstein focused on light waves, any one of which has a frequency f and a wavelength λ, and a universal speed of light $c = f\lambda$. But taken collectively, light waves (electromagnetic or EM waves) can have any value of f; there is no characteristic frequency for EM waves. But consider a quantum wave (or de Broglie wave) for an electron of mass m_e. This has a minimum frequency $f = m_e c^2/h$, which represents a characteristic frequency f_c (h = Planck's constant). Similarly, an electron has a characteristic length, the Compton wavelength $\lambda_c = c/f_c = h/m_e c$. So an electron is a natural clock and ruler which defines the local calibration of time and space. These change in a gravitational field, giving rise to the trajectories of general relativity. These trajectories may be computed using classical formalisms, without reference to abstract spacetime metrics.

Any system of physical units has three independent units (plus the fundamental electric charge e); for the standard metric system (SI) these are MKS, for meter, kilogram, and second. In the universal system described above, the natural units are f_c, λ_c, and h. Quantized spin provides the basis for h, and is Lorentz-invariant. Other parameters are defined in terms of these: $c = f_c \lambda_c$ and $m_e = hf_c/c^2$.

Within this picture, an electron is a real extended wave in real space, with quantized total spin. There is no intrinsic uncertainty, and no entanglement. Composites such as a proton or an atom are not quantum waves at all; they are bags of confined quantum waves, which inherit quantization from their constituents. Such a picture provides a simple, unified system with local reality all the way from atoms to galaxies and beyond.

Time is not a physical dimension; rather, it is simply a parameter that governs motion and change. Our most accurate clocks are atomic clocks, which inherit their properties from the fundamental electron clock. This defines the time of classical physics, which in turn defines our macroscopic clocks, as well as our biological and psychological clocks. Not all of these are quite so accurate, but they derive from the same physics. The direction of time follows simply from increasing microscopic entropy and macroscopic adaptation.

B. Quantum Computing is *not* the Future of Computing

There has recently been much attention to the potential of quantum computing, which promises the ability to perform computational tasks that are virtually impossible for any classical computer, regardless of the architecture. It has even been suggested that modern classical supercomputers will soon become obsolete due to "quantum supremacy". This essay proposes that classical computers with non-classical (neuromorphic) architectures may enable true artificial intelligence and even consciousness—why not quantum computers for this role? The reason is that I do not believe that the promised quantum computing is possible, on fundamental grounds (see [11]).

The key point is that the promised exponential enhancement in performance is based on the presence of quantum superposition and entanglement among N coupled quantum bits (or qubits), as implied by the orthodox Hilbert-space mathematical formalism for quantum states. I have questioned whether this mathematical formalism is correct [8, 10], and suggested some new experiments that can address this question. One can see the essential role of entanglement from the following simple argument. In any classical computing architecture, one enhances performance by the use of parallelism in hardware bits. If the hardware consists of N parallel bit slices, it can operate (ideally) N times as fast. In sharp contrast, for a quantum computing system, if there are N entangled qubits, this enables an effective computational parallelism by a factor of 2^N, which increases exponentially. For example, if $N = 300$, 2^N is greater than the number of atoms in the known universe. Clearly, no classical supercomputer with merely billions of bit slices can compete. The ability to obtain exponentially scaled equivalent performance from linear growth in hardware provides the entire motivation for quantum computing.

Such a claim of exponential performance is fantastic, in both senses of the word. Carl Sagan once said [20] that *"Extraordinary claims require extraordinary evidence"*. In fact, no such evidence exists, and people should be extremely skeptical. The primary reason why this has been accepted is that the orthodox theory of quantum mechanics, obscure and confusing though it may be, has been defended by most of the smartest minds in physics for decades. But early in the 20th century, both Albert Einstein and Erwin Schrödinger questioned the foundations of quantum mechanics. Recently, Weinberg [25] has also questioned these foundations.

Given the billions of dollars that are now being invested in quantum computing research by governments and corporations, I expect that this question will be settled within 20 years. I am not suggesting that quantum computing research is useless; on the contrary, it provides important insights into the isolation of nanoscale systems from environmental noise, which will be essential in the continued evolution of "classical" computer technology toward the atomic scale. But if I am correct, this will radically disrupt the orthodox understanding of quantum mechanics, leading to the adoption of a new quantum paradigm, with major long-term implications for the future of physics.

References

1. Dennett, D.C.: Why and how does consciousness seem the way it seems? In: Metzinger, T. (ed.) Open Mind (2015) (available online. See also TED Video 2003, "The Illusion of Consciousness")
2. Dick, P.K.: Do Androids Dream of Electric Sheep? Doubleday, New York (1968).(See also Wikipedia entry)
3. Dyson, G.: Turing's Cathedral: The Origins of the Digital Universe. Random House, NY (2012)
4. Einstein, A.: Quotation about Simplicity, see online. This seems to be a 1950 paraphrase of a more complicated statement by Einstein from 1934–see here for history. https://quoteinvestigator.com/2011/05/13/einstein-simple/
5. Gould, S.J.: The Panda's Thumb: More Reflections in Natural History. Norton, New York (1980) (See also Wikipedia entry)

6. Godfrey-Smith, P.: The mind of an octopus. In Scientific American Mind (1 Jan 2017) (available online. Also The Octopus, the Sea, and the Deep Origins of Consciousness, Farrar Strauss & Giroux, New York, 2016)
7. Hobson, A., et al.: Virtual reality & consciousness inference in dreaming. Front. Psych. (2014) (available online)
8. Kadin, A.M.: The Rise and Fall of Wave-Particle Duality (2012) (submitted to FQXi essay contest, available online). https://fqxi.org/community/forum/topic/1296
9. Kadin, A.M: Watching the Clock: Quantum Rotation and Relative Time (2013) (submitted to FQXi essay contest, available online). https://fqxi.org/community/forum/topic/1601
10. Kadin, A.M.: Remove the Blinders: How Mathematics Distorted the Development of Quantum theory (2015) (submitted to FQXi essay contest, available online). https://fqxi.org/community/forum/topic/2338
11. Kadin, A.M., Kaplan, S.B.: Proposed experiments to test the foundations of quantum computing. In: submitted to International Conference on Rebooting Computing (2016a) (available online). http://vixra.org/abs/1607.0105
12. Kadin, A.M.:Non-metric microscopic formulation of relativity. In: submitted to Gravity Research Foundation (2016b) (available online). http://vixra.org/abs/1603.0102
13. Kahneman, D.: Thinking, Fast and Slow. Farrar Strauss & Giroux, New York (2011). (See also Wikipedia entry)
14. Koestler, A.: The Ghost in the Machine. Macmillan, New York (1967) (See also Wikipedia entry)
15. Lahav, N., et al.: K-shell decomposition reveals hierarchical cortical organization of the human brain. New J. Phys. (2016) (available online, including embedded video)
16. Lewis-Krauss, G.: The great AI awakening. In: New York Times Magazine (14 Dec 2016) (available online)
17. Manzotti, R.: Machine consciousness: a modern approach. In: Natural Intelligence: The INNS Magazine, vol. 2, no. 1, p. 7, (July 2013) (available online)
18. Markham, J: How many computers to identify a cat? 16,000. In: New York Times (25 Jun 2012) (available online)
19. Monroe, D.: Neuromorphic computing gets ready for the (really) big time. Proc. ACM (2014) (available online)
20. Sagan, C: Cosmos: a personal voyage. PBS Television Series, 1990 Update, source of quote on "extraordinary claims" (1990) (See also Wikiquote entry)
21. Shakespeare, W.: As you like it. In: First Folio. Source of quotation "All the World's a stage …" (1623)
22. Turing, A.: Computing machinery and intelligence. Mind, vol. 256, p. 433, (1950) (available online)
23. Wanjek, C.: Sleep shrinks the brain–and that's a good thing. Sci. Am. (3 Feb 2017) (available online)
24. Weinberg, S.: To Explain the World: The Discovery of Modern Science. Harper-Collins, New York (2015)
25. Weinberg, S.: The trouble with quantum mechanics. In: New York Review of Books (19 Jan 2017) (available online)

Chapter 16
The Man in a Tailcoat

Tommaso Bolognesi

> As a practical matter we often end up describing what systems do in terms of purpose when
> this seems to us simpler than describing it in terms of mechanisms.
>
> Stephen Wolfram

16.1 Prologue

Yesterday around midnight I had an interesting conversation with an elegant man. With a top hat and a curious glass walking cane, he was standing on the parapet of a bridge, looking down at the dark river below his feet (Fig. 16.1).[1]

"What are your doing up there, Sir?", I politely asked. No answer. "Are you going to give up keeping your entropy low and far from thermodynamic equilibrium with the environment? Do you feel that your life has no purpose?" He turned his head and stared at me with a mix of surprise and irritation. "Well", I continued, "that's a bad idea. Purposes are overestimated concepts and an illusory business, both in life and elsewhere. Would you let me expand a little?"

I took his protracted silence as a consent, and proceeded.

---

[1] 'L'uomo in Frack' (The Man in a Tailcoat), also known as 'Vecchio Frack', is a poignant song by the italian singer and composer Domenico Modugno (1928–1994). Written in 1955, it describes the last hours of a mysterious character and is inspired by the true story of Raimondo Lanza di Trabia.

T. Bolognesi (✉)
Istituto di Scienza e Tecnologie dell'Informazione "A. Faedo", ISTI, CNR, Pisa, Italy
e-mail: t.bolognesi@isti.cnr.it

© Springer International Publishing AG, part of Springer Nature 2018
A. Aguirre et al. (eds.), *Wandering Towards a Goal*, The Frontiers Collection,
https://doi.org/10.1007/978-3-319-75726-1_16

Fig. 16.1 'L'uomo in Frack'
(painting by Dina Mosca)

16.2 Mechanisms and Goals

Garey and Johnson [4] define a *problem* as:

> a general question to be answered, usually possessing several *parameters*, or free variables, whose values are left unspecified. A problem is described by giving: (1) a general description of all its parameters, and (2) a statement of what properties the answer, or *solution*, is required to satisfy. An *instance* of a problem is obtained by specifying particular values for all the problem parameters.

Garey and Johnson are computer scientists. From their viewpoint (which may appear restrictive but is indeed fully general, as I'll clarify shortly) the solution of a problem is provided by an *algorithm*, which defines a step-by-step procedure, or computation, or process, or *mechanism*, for deriving from the problem instance (the input, or initial state) something (an output, or final state) that satisfies those required properties.

Terms like *purpose* and *goal*, that I use interchangeably, immediately fit into this picture: the *goal* of the *mechanism* is to satisfy the *properties* mentioned in point (2): you specify a goal by specifying those properties.

Here's an example of a mechanism *m* whose input is a tuple *in* of numbers, and whose details you don't need to grasp:

```
m[in_] :=  in //.
           {x_ _ _, a_?NumericQ, b_?NumericQ, y_ _ _} :> {x, b, a, y}
           /; b < a
```

And here is *m*'s goal *g* whose details might look more familiar (for simplicity I assume the numbers of tuple *in* to be all different):

$$g(in, out) =_{def} (set(in) = set(out)) \wedge (\forall j \in [1, n-1].out(j) < out(j+1)) \tag{16.1}$$

Mechanism *m* is written in the *Mathematica* programming language, and operates by swapping a pair of elements *a* and *b* whenever *a* precedes *b* in the tuple and $b < a$, until no such pair can be found. It is a rather naive sorting algorithm.

Goal *g* is written in first-order predicate calculus: it establishes that tuples *in* and *out* have the same elements and that the elements of *out* are sorted.

The equations that summarise Garey and Johnson's quote are then:

$$m(in) = out \tag{16.2}$$

$$g(in, out) = True \tag{16.3}$$

Both *m* and *g* define an input-output relation *R*. Mechanism *m* does it in a constructive way, 'by addition': given an *in*, it creates an associated *out* and adds pair (in, out) to *R*. Goal *g* does it in a declarative way and, in a sense, 'by subtraction': it puts all potential pairs in *R*, and then removes those that do not verify the requirement. Note that both construction and verification involve a computation, and in both cases the computation may diverge, since it is undecidable whether a generic *m* terminates on a generic *in*, and it is undecidable whether a generic predicate $g(in, out)$ is provable by the rules of first-order predicate calculus.

In principle, then, code *m* is as good as logic formula *g* for characterising an input-output relation. In practice, however, as the mechanisms become more and more complicated, a compact logic formula (whenever available!) is much more preferable for concisely characterising the process at hand, and for referring to it in human-to-human communication.

(In the darkness it is hard to decipher my interlocutor's expression. Puzzled?)

I guess you are tempted to criticise my approach as being too restrictive, since: (i) it illustrates *artificial* processes, like computer programs, but ignores *natural* processes; (ii) it focuses on mechanisms of a computational type, leaving out all the others. Fine! This gives me the opportunity to clarify why the computer scientist's viewpoint about mechanisms and goals is fully general and universal. That's because the whole physical world is itself algorithmic and carries out a gigantic computation, a relentless information processing activity that started with the Big Bang.

(At this point the Man in a Tailcoat gives a discreet cough.)

Thus, the correct way to look at a collision between two subatomic particles, a lightning, a biochemical reaction, the appearance of a new species, is in terms of a computing process. In this respect, the distinction between *natural* mechanisms, like these, and *artificial* mechanisms, like the computation of a sorting algorithm, is blurred. All we have is mechanisms—interacting, computational mechanisms all over the place!

"And goals? Do all mechanisms have a goal?" you may ask. (This seems to be a question of vital importance for The Man in a Tailcoat, tonight.)

For artificial systems, finding the correct (m, g) mechanism-goal pairing is usually trivial, since g is defined a priori by humans and the matching process m is designed by humans too, afterwards, and in a short time. This is the engineer's business.

But try to reverse the scenario and figure out g a posteriori, by just looking at the program code! This is the case with natural mechanisms. In nature no goal seems to be established a priori, by some Grand Architect, and we know today that the design of the various mechanisms, at least for the complex processes associated with life, is carried out by blind darwinian evolution, taking evolutionary times up to hundred-million years. Still the question remains: is it possible to find a concise formulation of a goal $g(in, out)$ for any natural mechanism m we observe, in the computation-oriented setting discussed above? This is the scientist's business, and for doing it our reasoning in terms of (in, out) pairs needs to be revised.

16.3   Compact Goals, Compressible Mechanisms, Decreasing Entropies

Many of the artificial mechanisms we build, e.g. sorting algorithms, are adequately characterised by their input-output behaviour, and by the compact $g(in, out)$ logical formulae conceived for defining their goals. When dealing with natural mechanisms this approach may become less effective, since the precise output of these processes is often hard to define. Rather, the mechanisms tend to run forever, so the $m(in) = out$ scenario must be revised in favour of a potentially endless $m(in)$ computation. For example, what is the final state of a planet orbiting around a star? On one hand, each individual step of the process is relevant and potentially eligible as an *intermediate* output of the computation, and we might ask what is the goal of the process *up to that point*.[2] On the other hand, when the output becomes a moving target, a reasonable alternative is to focus on the internal features of m in all its span, and replace the requirement of a clear and concise logical formulation of the (dynamic) input-output relation $g(in, out)$ by alternative 'lightness' requirements for the computation itself.

---

[2]In the context of the computational universe conjecture, the usual answer to the question "What is the *goal* of the universal computation?" is: "To compute the universe's own evolution" [6]. For each initial segment of the computation the 'goal' has been... just to push the universe up to that point. Tautological as it may sound, the answer has a genuine message: there is no shortcut to describing the evolution of the universe, other than going through it step-by-step.

For investigating alternative definitions of purposefulness, let us consider an example. In Fig. 16.2 we compare three mechanisms: the already introduced naive sorting algorithm and two elementary cellular automata (ECA)—ECA 110 and ECA 30. For ease of comparison we sampled the computations: the diagram for the sorting algorithm shows only every 100th row, while for the ECA's the sampling period is 8. Furthermore, rows are bit tuples of length 400 in all three cases. ECA's do process bit tuples. For the sorting algorithm, intermediate configurations would be tuples of integers: $(n_1, \ldots, n_{401})$. We turn them into bit-tuples $(b_1, \ldots, b_{400})$ by comparing adjacent numbers: $b_i = 0$ (resp. 1) when $n_i > n_{i+1}$ (resp. $n_i < n_{i+1}$).

Which of these mechanisms would we happily regard as purposeful?

A concisely expressed goal for the sorting mechanism (Fig. 16.2-left) was established a-priori by Eq. (16.1). Having turned numbers into bits, the goal component dealing with the comparisons of adjacent elements turns into the requirement for the output tuple to contain only 1's (black cells).

What about the computations of ECAs 110 and 30? As observed in [9] (p. 830), the first automaton *gives the impression* of being purposeful, the second does not. How can we substantiate these impressions?

When started from a random bit tuple, ECA 110 quickly develops digital particles, still perfectly visible with our periodic sampling (Fig. 16.2-center), that appear to interact and carry out themselves a computation. Indeed this automaton is Turing universal [2] and the motions of its particles can be exploited for simulating any (purposeful) algorithm, including sorting. Furthermore, with ECA 110 one can still somehow preserve the $g(in, out)$-oriented view. If *in* is some input bit tuple where a particle is already formed, and t is a predefined number of computation steps, one can predict the form of the output row *out*, due to the rectilinear motion of the particle: roughly, *out* will consist of the periodic background with the particle positioned at a distance s from its initial position, defined by a $s = vt$ type of formula. A careful study of particle collisions would allow to extend the approach to more complex scenarios. And we could again express the input-output relation $g(in, out)$ *concisely*.

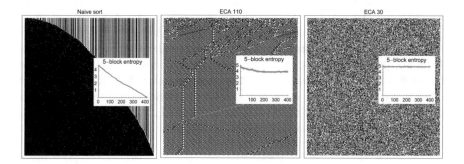

Fig. 16.2 5-block entropies for the naive sorting algorithm, ECA 110 and ECA 30 (sampled)

With ECA 30 (Fig. 16.2-right) the noise-like behaviour seems to prevent the identification of any meaningful goal.

Both for sorting and (to some extent) for ECA 110 we can find concise logical formulations of input-output relations. And both computations can be compressed. In the original sorting algorithm a pair of numbers (a, b) is considered at each step, and possibly swapped: one can compress the computation by considering, say, two appropriate pairs at a time. The compressibility of the ECA 110 computation is apparent from the sampled diagram, where particle interactions are preserved. Compressibility also derives from the highly regular structure of the background.

On the other hand, we could not associate a goal to ECA 30, and its computation is not compressible: to know the n-th row requires to compute all the preceding rows, one after the other. This circumstance may suggest to see compressibility as a necessary and sufficient condition for purposefulness. Would you agree?

(The Man in a Tailcoat remains silent and stares off into the distance. No, he does not seem to agree. Why?)

Oh, you are right! This would be a badly circular definition. We want to define purposefulness in terms of compressibility, but compressibility means reducing the length of the computation *without affecting its functionality*, that is, while preserving its original purpose! How can we safely compress a computation without knowing its purpose? How would we tell apart redundant from essential bits? Put it differently, if a purposeful mechanism must be compressible by definition, I could compress it until it becomes incompressible, without affecting its purpose: the result would be an incompressible, purposeful mechanism. A contradiction!

We can conclude that *we cannot take compressibility as an indication of purposefulness*. Well, thank you, my friend, you had a good point there! (but the Man in a Tailcoat, still standing on the parapet, does not react visibly to my gratifying remark.)

Let's now take a step forward (I mean... only metaforically!). There is another rigorously measurable quantity whose fluctuations might reveal the presence or absence of a goal: *entropy*.

The very peculiar entities that appear to manifest goals in this world—and intentions, consciousness, agency—have developed a great ability to maintain their entropy low, and lower than that of their environment. In a universe otherwise dominated by the 2nd principle of thermodynamics, this is indeed the mother of all goals—a tough job whose pursuit has been carved by darwinian evolution into the conscious and unconscious habits of living creatures, and one of the defining features of the emergent phenomenon of life. But just how early does this attitude emerge in the physical world? How much earlier than the appearance of the biosphere? What is the simplest mechanism, or system of open, interacting mechanisms in which localised entropy reduction may take place? (I extract a candle from my pocket, place it near the man's feet on the parapet, and light it up for illuminating the next figure.)

Let's consider the entropy fluctuations for the three processes of Fig. 16.2 (inset diagrams). For the two mechanisms that we recognise as purposeful—the sorting algorithm and ECA 110—entropy decreases; for the purposeless ECA 30, entropy remains stable to its maximum possible value 5.

The notion of entropy we have adopted here is Shannon entropy $-\sum_i p_i Log_2 p_i$, relative to bit-blocks of length 5. For each 400-bit row of the computations, the p_i's were obtained by counting the number of occurrences of the different bit-tuples of length 5—an arbitrarily chosen length. (Then, the 5-block entropy of a completely random bit row is 5 because the 32 distinct 5-blocks are equiprobable, with $p_i = 2^{-5}$.)

Expectedly, the decrease of entropy in the first two examples of Fig. 16.2 corresponds to a growth of perceived order and regularity. In a computational universe consisting of all conceivable mechanisms—random algorithms operating on random inputs, or monkeys typing at random on computer keyboards—it is well known that the outputs follow the universal a priori probability, or Solomonoff–Levin algorithmic probability [5], with a more pronounced bias towards regularity (Fig. 16.3-lower) than in a truly random universe—monkeys typing on plain old typewriters (Fig. 16.3-upper). Thus, if the conceptual equation linking goals to local entropy reduction is correct, we could say that the existence of mechanisms with a goal in the computational universe does not come as a complete surprise...

(In fact, my interlocutor does not look surprised at all, at the moment. Bored?)

Ok, I grant you that this may sound too abstract. In particular, the eight patterns in the lower part of Fig. 16.3 are the outputs of eight mechanisms that have proceeded in complete independence from one another for 80,000 steps. We could do better by letting mechanisms cooperate.

A closed physical system—one unable to interact with its environment—cannot hope to invert the natural tendency of its entropy to grow. An open system can. Our Earth, for example, is an open system that interacts with outer space. As observed in [7], the energy it receives from the sun during the day is roughly the same it returns to dark space at night, but, due to the large temperature gap between the sun and dark space, the incoming photons are individually more energetic, thus less numerous, than the outgoing, less energetic photons. Less incoming than outgoing photons implies less incoming than outgoing entropy. This is the way our planet manages to keep order—low entropy—in the biosphere.

Consider now a toy computational universe consisting of the seven ECA's {41, 48, 54, 98, 102, 158, 206} acting independently from one another, like the eight Turing machines of Fig. 16.3-lower row. The entropy levels attained by these

Fig. 16.3 Upper: random typing on eight paper sheets, using a three-character (three-color) typing machine. Lower: results of 80,000-step computations of eight randomly chosen 2-dimensional, 3-state, 3-color Turing machines, each running with the corresponding sheet above as input. (From [1])

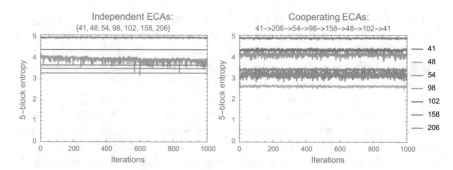

Fig. 16.4 Entropies of seven ECAs acting independently (left) and in cooperation (right)

automata are shown in the left diagram of Fig. 16.4. In the simulations we have run each ECA for 1 million steps, starting from a random 600-bit array, and have computed the 5-block entropies of every 1000-th step, obtaining a plot of 1000 entropy values for each automaton. In spite of the wider oscillations of ECA's 54 and 102, each ECA settles quickly to its own peculiar level, ranging between value 3.3 achieved by ECA 98 and a value a little below 5 achieved by ECA 102.

What happens if we let the ECA's cooperate? In this slightly less naive universe we have arranged the automata in a cycle: $41 \rightarrow 206 \rightarrow 54 \rightarrow 98 \rightarrow 158 \rightarrow 48 \rightarrow 102 \rightarrow 41$, so that the output of one is fed into the next. Each automaton runs for 1000 steps before offering its output, and a total of 1000 such iterations are performed, as in the previous simulation.

The interesting effect of cooperation is to spread entropy values in a wider range. In particular: the values of ECA's 102 and 41 are not affected; those of ECA's 54 and 98 slightly increase; those of ECA's 48, 158 and 206 *decrease*. The entropy decrease of ECA 48 is remarkably close to one unit!

In 'real' living organisms one can in principle establish an entropy threshold above which the organism is dead (an idea which is also central in Carlo Rovelli's essay, found in the present collection [8]). Assume we decided: (i) to attribute an alive/dead status to the seven inhabitants of our toy universe, (ii) to do it just in terms of their entropies, and (ii) to set their life-to-death entropy threshold at value 3. Under these assumptions, we should conclude that none of the automata is alive in the scenario of independent systems, while one of them—ECA 48—would come to life when they cooperate. You may certainly dismiss these bizarre assumptions as a blatant over-simplification of the multiple requirements behind the complex phenomenon of life. I would agree. This is meant to be just a proof of concept. But the concept it proves is interesting: the game of local entropy reduction, that we recognise as a crucial ability of living organisms, may well be played also among formal, abstract algorithms that are normally considered as totally extraneous to the mechanisms of biology.

(The starry night must have reached its coldest peak. As dawn approaches, my silent interlocutor keeps staring at the river below us, as if dangerously attracted by the highly entropic flow of low-frequency photons that keep escaping the planet.)

Ok, my friend, you are right: I started our conversation by mentioning the illusory nature of goals and purposes, but have not yet justified that claim. Let me do it now.

16.4   Goals, Emergence and Abstraction

Simply stated, my point is that mechanisms exist in nature as the fundamental building blocks of External Reality: they enjoy the most respected ontological status. Goals don't. Goals are only a convenient product of human knowledge, an epistemological device, a mental construction meant to offer practical representations of the mechanisms we observe *in the upper levels of the universal architecture of emergence*, those that we inhabit ourselves. Understanding those mechanisms in terms of compactly expressible goals has indeed given a formidable evolutionary advantage to our species, much as the development of human languages did, and this explains why goal-oriented reasoning is so widely spread and why we attribute so much importance to it.

I do insist that the formulation of concise goals starts to become possible as we proceed upward in the hierarchy of emergence. Up in the highest levels, in the biosphere, the illusory nature of goals is well understood and accepted: they are just a *powerful narrative trick* for describing, a-posteriori, features of mechanisms that darwinian evolution developed without any a-priori blueprint. We are so much accustomed to goal-oriented storytelling that we sometimes attribute intentions even to computers, cars, vacuum cleaners, when we describe their behaviours in anthropomorphic terms!

But what about the existence of goals as we proceed *downward* and look at very simple computational mechanisms? Recall the formulation of mechanisms and goals in terms of, respectively, an algorithm $m(in)$ and a logic formula $g(in, out)$, and the requirement that g be a *compact* logical predicate that can be more readily understood and communicated than the description of mechanism m (the tradeoff being, in general, that g defines the input-output relation in a non-constructive way).

Let us again consider ECA 110. What made us think that this elementary cellular automaton has a goal is, ultimately, the emergence of particles. As already mentioned, at least in the case of a single particle we can conceive compact formulations of input-output relations—involving the initial and final position of the particle—essentially in terms of the laws of rectilinear motion, while completely ignoring the corresponding mechanism. The mechanism, of course, consists in the *interplay of the multitude of cells* that make up the automaton, all characterised by the boolean function f_{110}: $\{0, 1\}^3 \rightarrow \{0, 1\}$. This function defines the next value b' of a cell in terms of the current value b of the cell itself and of the values a and c of its left and right neighbors: $b' = f_{110}(a, b, c)$. Thus, when we focus on particles, at the viewpoint of emergent

Level 1, we can successfully distinguish between goal (say, a particle's target) and mechanism (collective, iterated applications of function f_{110}), as in many biological processes.

Can we do the same when moving down to Level 0? In other words, can we distinguish between a goal and a mechanism when looking at a *single step* of an *individual* ECA 110 cell—a *single* application of the defining function of that automaton?

Function f_{110} is defined as follows:

$$f_{110}(a, b, c) =_{def} Xor[Or[a, b], And[a, b, c]] \qquad (16.4)$$

You may suggest that Eq. (16.4) describes a *mechanism* since it is expressed in terms of more elementary boolean functions (*Xor, Or, And*) whose 'execution' can be carried out in a sequence of computational steps, as in a program.

How about a corresponding *goal*? How can we *concisely* characterise the input-output relation implemented by a boolean function such as f_{110}? In general, this is done by a truth table, which consists in the enumeration of all input-output pairs—in this case, of form $(a, b, c) \rightarrow b'$:

$$\{(1, 1, 1) \rightarrow 0, (1, 1, 0) \rightarrow 1, (1, 0, 1) \rightarrow 1, (1, 0, 0) \rightarrow 0,$$
$$(0, 1, 1) \rightarrow 1, (0, 1, 0) \rightarrow 1, (0, 0, 1) \rightarrow 1, (0, 0, 0) \rightarrow 0\}. \qquad (16.5)$$

But this brute-force solution—which is anything but compact—is not what we need, since we are looking for a concise *logical formula* as the one in Eq. (16.1).

To summarize: at Level 1 (particles), I can conceive independent descriptions for the goal—dealing with the properties of the emergent particle dynamics—and the mechanism—dealing with the cooperative behaviour of a multitude of cells, each behaving according to function f_{110}. At Level 0 (individual cell performing one step) I can't. A single step of a single cell turns out to be too simple a phenomenon for admitting a double characterisation in terms of a mechanism *and* a goal: no such distinction seems possible.

So, the lesson from cellular automata is that goals prosper only when emergent, collective phenomena start to manifest themselves: at the *fundamental*, ground level, they completely loose grip! Therefore, if you are into *foundational questions* (as you seem to be), a lack of goals is no big deal.

(Too much of a stretch? Indeed, there's no sign of agreement on the face of my silent interlocutor... But no disagreement either! This encourages me to elaborate my last claim, trying to widen its span.)

With cellular automata—at least with some of them—the formulation of a goal is made possible by the existence of a certain degree of complexity in their mechanism—a mechanism involving a whole collection of interacting agents. How general is this observation about goals?

For seeing this, let me first introduce a new keyword, one that allows us to look at goals under a new perspective: *abstraction*. We may say that a goal *abstracts away* the complexity of the mechanism to which it is paired. Let's now move from CA's to generic algorithms. The goal expressed in Eq. (16.1) for sorting algorithms can be certainly seen as an *abstraction* of the specific sorting mechanism I provided at the beginning of our conversation, since it does not specify the details of how the algorithm should operate. As a consequence, the goal-mechanism relation appears as a one-to-many relation: several different mechanisms—e.g. *BubbleSort, QuickSort, MergeSort, HeapSort*, etc.—can implement the same goal—e.g. sorting.

Abstraction is certainly relevant also w.r.t. the other topic we discussed earlier, when we related goals with entropy decrease. There is no way to talk about entropy without invoking *emergence* and a *one-to-many* relation of *abstraction*. Indeed, for entropy to be definable you need: (i) a collection Ψ of entities—e.g. gas molecules, (ii) a usually huge space of micro-states of Ψ, ranged over by variable x, each describing *in detail* (i.e., entity by entity) a configuration of Ψ, and (iii) a set of macro-states of Ψ, ranged over by variable X, each characterising a different equivalence class $C(X) = \{x_1, x_2, \ldots, x_n, \ldots\}$ of micro-states. A macro-state X *abstracts away* the details of the micro-states in its equivalence class $C(X)$: one-(macro-state)-*to-many*-(micro-states). Then, the entropy $S(X)$ of macro-state X is proportional to the *log* of the size $W(X) = |C(X)|$ (counting micro-states) of the corresponding class:

$$S(X) \propto log(W(X)) \tag{16.6}$$

(You see the importance of the collective—of the global picture? How could we talk about the entropy of a single particle in an empty volume of space?)

Now, a mechanism is concerned with micro-states, a goal with macro-states. Macro-states, thus goals, are, again, *emergent*: they appear only under a global perspective, when our observational apparatus is not good enough to discriminate among equivalent micro-states, but is only sensitive to macro-states, e.g. it only detects temperature, volume, pressure. In conclusion: goals are the illusory consequence of the weakness of our sight!

"Temperature, volume, pressure? None of these macro-variables was mentioned when you discussed entropy decrease in cellular automata! You only monitored the fluctuations of a somewhat obscure '5-block entropy' for some seven automata. That does not seem to be the entropy defined in Eq. (16.6). What is macro-state X in that context, what is equivalence class $C(X)$, and what its size W, so that one can consider its *log*?". (Not that the Man in a Tailcoat asked all this, but he may be silently wondering... So in these last minutes before dawn I set to clarify this final point.)

16.5 Shannon Block-Entropy Versus Boltzmann Entropy

For fixing ideas, and for simplicity, let us start with blocks of length 1, while still considering ECA rows of length 600, as adopted for the experiment documented in Fig. 16.4. For computing the 1-block entropy of some ECA row r—a tuple of 600 bits—we start by counting the number of occurrences in r of the different blocks of length 1, namely blocks (0) and (1)—meaning that we plainly count the 0's and 1's. We keep the count as a pair of integers (n_0, n_1), with the obvious meaning of the two variables.

The block entropy is based on Shannon's definition, which is indeed given in terms of a probability distribution d for the 1-blocks, thus a pair of probabilities for bits 0 and 1: $d = (p_0, p_1)$. But we are not given such a distribution in advance, we only have the bit count. The most probable distribution that reflects bit count (n_0, n_1) is $d = (n_0/600, n_1/600)$: these are the probabilities that we use in Eq. (16.8) below for obtaining the 1-block entropy.

$$H(X) = H((n_0, n_1)) = - \sum_{i \in \{0,1\}} (n_i/600) log_2(n_i/600) \qquad (16.7)$$

Note that this is the expected amount of information carried by a generic 1-block, i.e. obtained from reading a generic cell. Since there are 600 1-blocks, we can also express the 1-block entropy of the *whole* row as follows:

$$H^{600}((n_0, n_1)) = 600 * H((n_0, n_1)) \qquad (16.8)$$

The above definition of block-entropy reflects an *intensional* approach, since it focuses on the internals of the 600-bit row. How about the *extensional* approach that seems to inspire the definition of S in Eq. (16.6)? How about micro-states, equivalence classes, macro-states, class sizes? Well, they are all here! A micro-state x is a specific 600-tuple of bits. A macro-state X is the bit-count *pair* (n_0, n_1): we assume that our limited sight only allows us to detect the proportion of 0's and 1's of a micro-state, not the way bits are arranged in the tuple. The equivalence class $C(X)$, for $X = (n_0, n_1)$, is then the set of all 600-tuples of bits that share that same count of 0's and 1's. The size of this set is easily determined:

$$W(X) = W((n_0, n_1)) = \frac{(n_0 + n_1)!}{n_0! n_1!} = \frac{600!}{n_0! n_1!} \qquad (16.9)$$

We have now all ingredients for applying Eq. (16.6):

$$S(X) = S((n_0, n_1)) \propto log\left(\frac{600!}{n_0! n_1!}\right) \qquad (16.10)$$

Note that Eq. (16.6) is valid when all the equivalent micro-states of class $C(X)$ have equal probability.[3] This is indeed the case, no matter which distribution $d = (p_0, p_1)$ we consider, since the probability of *any* $(n_0 + n_1)$-tuple with *fixed* bit count (n_0, n_1) is $p_0^{n_0} p_1^{n_1}$, where p_i is derived from n_i as explained above.

Let us now compare numerically the two 1-block entropies—the intensional $H^{600}(X)$ and the extensional $S(X)$. This is done in Fig. 16.5-left. The values of the

---

[3] When each micro-state c_i in $C(X)$ has its own probability p_i, the entropy is $S \propto - \sum_i p_i log(p_i)$.

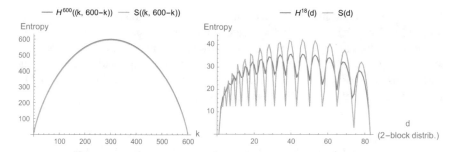

Fig. 16.5 Left: H^{600} entropy and S entropy for 600-bit rows with all possible 1-block distributions, corresponding to all possible bit counts $(n_0, n_1) = (k, 600 - k)$, with k ranging in 0–600. Right: H^{18} entropy and S entropy (the latter is computed with base-2 logarithm, and amplified by an arbitrary factor 3) for all possible 2-block distributions d, corresponding to all possible 2-block counts $(n_{(0,0)}, n_{(0,1)}, n_{(1,0)}, n_{(1,1)})$, with $n_{(0,0)} + n_{(0,1)} + n_{(1,0)} + n_{(1,1)} = 18$

two entropies have been computed for all possible bit counts (n_0, n_1), represented along the x-axis as pairs $(k, 600 - k)$, with $k = 0, 1, \ldots, 600$. Note that formula (16.10) for the S entropy involves an unspecified proportionality factor: for the plot in Fig. 16.5-left we have used the base-2 logarithm, and proportionality factor 1. With this setting the two entropies are almost identical, although their difference becomes more appreciable for smaller row lengths. An analytic account for the closeness of these two measures is given in the appendix.

In Fig. 16.5-right we plot the two entropies H and S relative to the counts of 2-blocks in bit rows of length 18. When dealing with 2-blocks, each row instance is characterised by a quadruple $d = (n_{(0,0)}, n_{(0,1)}, n_{(1,0)}, n_{(1,1)})$ of integers, sloppily called itself a 'distribution', indicating the counts of the four types of 2-blocks occurring in the row. While with 1-blocks the number of different rows matching a given count pair (n_0, n_1) is readily computed by Eq. (16.9), with 2-blocks a similar count is less trivial, and as a workaround one can explicitly enumerate all the 2^L rows of length L, each time detecting the associated distribution d and keeping a global count of the occurrences of the latter. The drastic reduction in row length is due to this computational bottleneck of explicit row enumeration.

In Fig. 16.5-right, for the S entropy we have used the base-2 logarithm and an arbitrary proportionality factor 3, for a convenient visual comparison. We still detect a strong correspondence between the Shannon block-entropy H^{18} and entropy S, which—I may have forgotten to mention—is due to Boltzmann...

16.6   Epilogue

For reasons that are still obscure to me, pronouncing the famous physicist's name must have triggered a deadly flow of neural micro-state changes in the head of the man, with visible consequences on the macro-variables characterising both his facial expression and his precarious equilibrium on the parapet. "Are you perhaps a theoretical physicist yourself?" I asked, while abruptly grabbing and pulling the lower end of his walking stick in an attempt to stop an unwholesome action. Smooth and slippery as it was, the crystal cane remained firm in my hand, not in his. The end of this story of local entropy growth—too sad to be reported here—is told by the final verses of the song 'The Man in a Tailcoat'.

Appendix—H Versus S via Log-Factorial Approximation

In general, we can rewrite the 1-block entropy $H^{(n_0+n_1)}$ of Eq. (16.8) (where $n_0 + n_1 = 600$) as:

$$H^{n_0+n_1}((n_0, n_1)) = (n_0 + n_1)\left(-\frac{n_0}{n_0 + n_1}log_2\frac{n_0}{n_0 + n_1} - \frac{n_1}{n_0 + n_1}log_2\frac{n_1}{n_0 + n_1}\right)$$
$$= (n_0 + n_1)log_2(n_0 + n_1) - n_0 log_2(n_0) - n_1 log_2(n_1) \quad (16.11)$$

Using base-2 logarithms and proportionality factor 1, the Boltzmann entropy of Eq. (16.10), relative to tuples with bit count (n_0, n_1), becomes:

$$S((n_0, n_1)) = log_2\frac{(n_0 + n_1)!}{n_0!n_1!}$$
$$= log_2 e * (log((n_0 + n_1)!) - log(n_0!) - log(n_1!)) \quad (16.12)$$

where $log()$ denotes the base e natural logarithm. By using the following approximation of the log-factorial function [3]:

$$log(n!) \approx \left(n + \frac{1}{2}\right)log(n + 1) - (n + 1) + \frac{1}{2}log(2\pi) + \frac{1}{12(n + 1)} \quad (16.13)$$

we further develop Eq. (16.12) as follows:

$$S((n_0, n_1)) \approx \left(n_0 + n_1 + \frac{1}{2}\right)log_2(n_0 + n_1 + 1) - \left(n_0 + \frac{1}{2}\right)log_2(n_0 + 1)$$
$$- \left(n_1 + \frac{1}{2}\right)log_2(n_1 + 1) + SS(n_0, n_1) \quad (16.14)$$

where:

$$SS(n_0, n_1) = log_2(e) - \frac{1}{2}log_2(2\pi) + \frac{log_2(e)}{12}\left(\frac{1}{n_0 + n_1 + 1} - \frac{1}{n_0 + 1} - \frac{1}{n_1 + 1}\right)$$

$$= 0.117 + 0.120\left(\frac{1}{n_0 + n_1 + 1} - \frac{1}{n_0 + 1} - \frac{1}{n_1 + 1}\right) \quad (16.15)$$

Clearly the contribution of $SS(n_0, n_1)$ is negligible, while the first three terms of the approximated $S((n_0, n_1))$ of Eq. (16.14) parallel closely the three terms in the expansion of $H^{n_0+n_1}$ provided by Eq. (16.11). This explains the closeness of the two plots in Fig. 16.5-left.

References

1. Bolognesi, T.: Spacetime computing: towards algorithmic causal sets with special-relativistic properties. In: Adamatzky, A. (ed.) Advances in Unconventional Computing - vol. 1: Theory, vol. 22. Emergence, Complexity and Computation, pp. 267–304. Springer, Berlin (2017). Chap. 12
2. Cook, M.: Universality in elementary cellular automata. Complex Syst. **15**, 1–40 (2004)
3. Cook, J.D.: How to compute log factorial. Posted 16 August 2010. https://www.johndcook.com/blog/2010/08/16/how-to-compute-log-factorial/
4. Garey, M.R., Johnson, D.S.: Computers and Intractability—A Guide to the Theory of NP-Completeness. Freeman, San Francisco (1979)
5. Hutter, M., Legg, S., Vitanyi, P.M.B.: Algorithmic probability. Scholarpedia **2**(8), 2572 (2007). (Revision 151509)
6. Lloyd, S.: Universe as quantum computer. Complexity **3**(1), 32–35 (1997)
7. Penrose, R.: Cycles of Time. The Bodley Head, London (2010)
8. Rovelli, C.: Meaning and intentionality = information + evolution. In: Aguirre, A., Foster, B., Merali, Z. (eds.) Wandering Towards a Goal - How Can Mindless Mathematical Laws Give Rise to Aims and Intentions. Springer, Berlin (2018). (This volume)
9. Wolfram, S.: A New Kind of Science. Wolfram Media, Inc., Champaign (2002)

Chapter 17
The Tablet of the Metalaw

Cristinel Stoica

> I don't believe in empirical science. I only believe in a priori truths.
>
> Kurt Gödel

17.1 The Tablet of the Law

Many physicists share the dream that sooner or later we will know the fundamental physical laws, the equations describing them, and that they will be unified in a single theory which fits on a t-shirt, or perhaps on a stone tablet—*the tablet of the law*.

Maybe it is just a dream, at least this is still a matter of belief. There is no known reason to be true—why would the universe be fundamentally simple, completely describable by a finite set of laws? Why would we be capable to understand them all? Maybe there are truths that simply cannot be discovered, or tested, or are too complex to be understood by our minds, even if we evolve indefinitely. And I don't mean that there are magical, supernatural phenomena that science can't explain, I simply mean that there is no reason to believe that a subset of the universe is able to test through experiments all aspects of the universe, neither to imagine a rigorous and precise model of everything. By contrary, I think that the idea that we are able to understand and test everything is based on magic, as if the universe would intentionally allow its code to be comprehensible and testable by human beings.

Einstein said

> I believe in Spinoza's God, Who reveals Himself in the lawful harmony of the world, not in a God Who concerns Himself with the fate and the doings of mankind.
>
> *Einstein to Rabbi Herbert Goldstein, 1929.*

C. Stoica (✉)
National Institute of Physics and Nuclear Engineering—Horia Hulubei,
Bucharest, Romania
e-mail: holotronix@gmail.com

© Springer International Publishing AG, part of Springer Nature 2018
A. Aguirre et al. (eds.), *Wandering Towards a Goal*, The Frontiers Collection,
https://doi.org/10.1007/978-3-319-75726-1_17

Fig. 17.1 The tablet of the
law, as known today

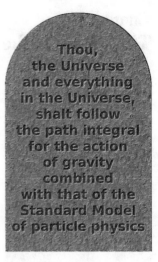

Thou,
the Universe
and everything
in the Universe,
shalt follow
the path integral
for the action
of gravity
combined
with that of the
Standard Model
of particle physics

Why would Spinoza's and Einstein's God, while not being concerned with us,
choose to reveal to us His most intimate thoughts?

This optimism is fueled by the foundation of the entire physics on a few laws we
know, of *general relativity, quantum theory*, and the *Standard Model* of particles,
which indeed fit on a tablet (Fig. 17.1). This makes us hope that the solutions of
puzzles like *dark matter, dark energy*, and *quantum gravity*, will still fit on the tablet.
I share this dream too—I even try to solve some problems of these theories, to make
them compatible for a future unification: the *singularities* of general relativity [1],
the problem of the *wavefunction collapse* in quantum mechanics [2, 3], and a unified
model of the Standard Model particles and gauge symmetries [4].[1]

Although our theories describe so concisely so much about the universe, the puzzle
is not completely solved. The known matter seems to account for less than 5% of
the mass-energy in the universe. The discrepancies between the predicted and the
observed rotational patterns of the galaxies suggest that either there are more particles
than we know, with strange new properties which make them not directly observable
(*dark matter*), or that gravity as described by general relativity should be modified. If
the *Lagrangian* of general relativity is not the Ricci curvature R, but a function $f(R)$,
it may be possible that we can never know that function f. Maybe we can determine
good estimates of the first terms of its power series, but what if the series is infinite
and doesn't follow a rule? Another missing piece of the puzzle is the compatibility
between quantum theory and gravity, which may require infinitely many coupling
constants. And take into account that what we know is based on observations made
in an infinitesimal zone of the universe, for a very brief period of time compared with
the age of the universe, so we can never exclude the possibility of new surprises. But

---

[1]The model proposed in [4] unifies into a simple algebra leptons and quarks of a generation, as
well as the gauge symmetries, in a minimal way, without predicting new particles, forces, or proton
decay.

at least we can hope that there is a small number of laws to add to the list of those we know so far, and even if they depend on an infinite number of parameters, that a finite number of them are relevant.

However, no matter how much we would bet on the unification, we have to take into account the possibility that simply there are pieces of the puzzle that we will never even find, let alone to fit them properly into the big picture. But this should never stop us trying at least to solve it as much as possible.

17.2 An Abstract Painting of the Universe

A useful picture of how a theory of the universe works is given by *dynamical systems*. All possible states of a system are parametrized by some values, and are collected into a space whose dimensions are those parameters. For example, in classical mechanics, to represent the state of a system of n point particles and how will it evolve, we need to know the position and the velocity of each particle, so $(3+3)n$ parameters, giving $6n$ dimensions. For classical fields, which can take distinct values at each point in space, like the electromagnetic field, we need an infinite number of parameters, and the same is true for continuum mechanics. The resulting space is called *phase space*. In quantum mechanics the number of dimensions is infinite even for a single particle, and *the state space* is a complex vector space. Then, we need a rule to specify how the system changes from one state to the next—usually an equation of motion. The evolution of such a system, no matter how many parameters define it, is a trajectory in the phase or state space.

Quantum mechanics can be obtained from classical mechanics by *quantization*, which basically consists in replacing the points in the phase space with complex functions defined on the entire phase space, followed by some more abstract steps [5]. But this picture can also be made into a dynamical system, by collecting into a space all these complex functions. The resulting space gains a structure of Hilbert space, and the dynamical law is the Schrödinger equation, or some of its relativistic versions.

Even if the rule is not deterministic, the dynamical system picture still works, if we describe the potential evolution of the system as forking into more possibilities (Fig. 17.2).

What's beautiful about dynamical systems is that it seems that any kind of theory attempting to describe the universe fits the pattern. Even the theory of relativity, where time is relative, can be described as a dynamical system, after slicing the spacetime into slices of equal time [6]. Even computation is described like this, including cellular automata [7]. Deterministic and indeterministic, continuous or discrete systems or theories, they all follow the pattern. So even if we don't know completely the fundamental laws, at least we know that they behave like dynamical systems. Maybe the task of physics is to find the complete description of the world as a dynamical system, and of science to explain all phenomena by reducing them to this.

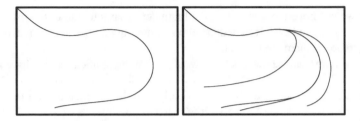

Fig. 17.2 Deterministic versus indeterministic dynamical systems

The abstract painting of the universe is a single point, but the canvas is likely infinite-dimensional. Its laws are curvy strokes on that canvas. The tablet of the law may very well contain just the equations of these lines.

17.3   The Structure of the Point

In the picture given by dynamical systems, how can a point contain complete information about the entire state of the system, in particular of the universe? All this information is not encoded in the point, but in its location in the state space. The state space is not a structureless collection of points, each place represents a different state.

Often we need to know more details about the state of the system, for example if we are interested in subsystems, or in the *universality of laws*—their independence of position and time, or in invariance. We may want to describe interactions and causal relations between parts, entanglement, self-organization etc. To do this, we have to go back from the representation of the state as a point to its description as a collection of particles or fields, or whatever structure the state has. In [8] I proposed a mathematical structure that is as general or even more general than that of dynamical systems, and at the same time deals more directly with the detailed structure of the state, with invariance of the laws, causality, emergence, metaphysical problems etc. But for the present discussion, dynamical systems are enough.

17.4   Zoom-Dependent Reality

Our direct experience seems to conform to the laws of Newtonian mechanics. But if we experiment with small distances, quantum theory becomes manifest through strange phenomena like quantum superposition, complementarity, uncertainty, and entanglement. At the most fundamental level we know there are quantum fluctuations, and as we zoom out we encounter particles, atoms, molecules, cells, tissues, biological organisms, societies, planets, galaxies and so on. For higher levels the

Fig. 17.3 Multiple levels of
a dynamical system

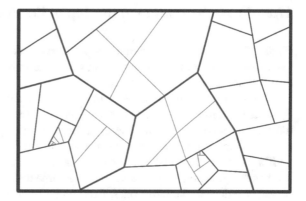

governing laws seem to be completely different from those at the lowest level. It seems that reality depends on the zoom level (Fig. 17.3).

Even at the higher levels, the picture can be understood using dynamical systems. Take for example molecules, they take part in chemical reactions, and the laws of chemistry describe how a system of molecules changes in time. Or think at the human-size level, where Newtonian mechanics is accurate enough. We can describe this classical level as a dynamical system too, which is nothing like the one used for the microscopic level. On the same canvas there is a classical painting, and if we zoom in, at each zoom level there is a completely different painting. Eventually we find a quantum painting.

There is a connection between all paintings visible on the same canvas—what appears structureless at one level, has a rich structure at the lower levels, as if each level is nested inside the next higher level. The highest levels result from the lowest level by ignoring details, resulting in a *coarse graining* of the state space. This is a process of abstraction, captured mathematically as follows. Two states of a low level system or subsystem are considered equivalent at a higher level if they cannot be distinguished at that level. This relation of equivalence partitions the low level phase space into equivalence classes, or *coarse graining* regions. Mathematically speaking, the inability to distinguish, the "confusion", is the equivalence relation on the phase space. The higher level may be described as a dynamical system, which is related to the lower level by a *forgetful functor*. Then, we identify each such region to a point and construct the phase space of the higher level.

In general the evolution law of the lowest level leads at a higher level to an approximate or statistical law, because the details of the lower level which we ignored can become relevant at the higher level. An example of statistical laws resulting from a deterministic lower level is the emergence of thermodynamics from classical mechanics. An example when the higher level's law is approximate is the emergence of the classical level of reality from the underlying quantum level.

17.5   From the Bottom to the Top of the Pyramid

Finding the complete fundamental laws is not easy, and even if we find a description that explains everything we know so far, we can never be sure that new experimental data will not require us to change it.

Suppose anyway that at some point we will find them. Suppose that the God of Spinoza and Einstein decides to handle to us the tablet of the law. Then what? Is this the end of science? Is this the Theory of Everything? By knowing the fundamental laws of nature, will we be able to deduce everything from them?

If a set of fundamental laws governs the world at the subatomic level, then it must govern it at all levels. It is natural to assume that higher levels of organization of matter reduce to the lower ones, from social behavior to psychology to biology to chemistry to atomic physics to the fundamental laws. The reductionist point of view works so well in constructing explanations and offering predictions. Maybe we don't always feel comfortable with the idea that all we are is just a bunch of mindless interacting particles. But let's forget about what makes us feel comfortable, or of the ethical implications of reductionism. Let's ask science itself if it's possible to reduce everything to a set of fundamental principles or matter constituents.

Science is a human activity, and humans are limited. We can only process finite amounts of information, and make only deductions or mathematical proofs of finite length. But the laws of the universe may not have this limitation. The universe knows how to build atoms, but we only know how to describe without approximations the simplest one (well, actually even to calculate the electron wavefunction for the Hydrogen atom we approximate the proton with a point). The universe knows how to evolve, but we can only know a finite amount of information even about the state of a small subsystem. And out of this, we can only calculate the evolution of the system with approximation. We would have the same limitations even if it would turn out that the phase space of the universe is discrete. Number theory is discrete, but Gödel's incompleteness theorem prevents us to know everything about it—Gödel's incompleteness theorem limits our possibility to deduce every consequence of a set of axioms by finite proofs, but it doesn't prevent us from finding the mathematical structure isomorphic to the universe [9]. Even computers have limitations—computer algorithms are finite, but Turing proved that there is no algorithm able to decide whether an arbitrary algorithm will halt or run forever.

Note that the way no-go results like Gödel's and Turing's limit the power of reductionism is purely about computability and the possibility to give finite-length proofs. In fact, the fundamental laws never cease to be true at the higher levels, and the high-level behavior is determined by the low-level detail, but it cannot be deduced from it. The higher levels have their own rules, emergent from the fundamental laws, implemented on top of them, but the opposite is to some extent also true, as we will see that it is the case for the relation between the quantum and the classical levels of reality.

Fig. 17.4 Incompleteness versus reductionism. The possibility to reduce the higher levels to the lower level is bounded by our limits, but also the limits of computation and mathematical proof

Because of these limitations we have, it is possible that some phenomena are not comprehensible to us (Fig. 17.4). But I think that one should choose, whenever possible, the reductionist explanation.

However, there is a bigger limitation of reductionism, manifest in quantum mechanics (see Sect. 17.8).

17.6   Floating Levels of the Pyramid

If we can never be sure what is written on the tablet of the law, how is science even possible? How can we understand a complex process, if we don't know its fundamental constituents and what they do?

This is possible to a surprising degree, and we are doing it for long time. Through observations and experiment, along centuries, we identified objects—relatively stable structures—and we guessed some approximate rules describing their behavior. We did this by ignoring the details, through abstraction.

Abstraction came naturally, because there are patterns that we could observe, and we didn't have access to the details, or we ignored them because they are too complex. If Galileo used to throw rocks to study gravity, he could only ignore the detailed shape and the constituents of the rocks, approximating them as points. Such idealizations allowed the discovery of mechanics by him, Newton, and others. And even now, when we know better what makes those rocks, we are still using abstraction, both because our limited knowledge and understanding, and because this allows us to still get useful results.

We can think of a rock, or of the laws of thermodynamics, as things that emerge from the fundamental laws. But what's beautiful is that we can ignore the details, and derive conclusions which remain true even if we change the detailed explanation. Heat is motion of the molecules composing a large system, no matter what those molecules are—they may be classical balls, or quantum systems (and by quantum we can mean anything that our theory and its interpretation says it is). What's important

Fig. 17.5 Dynamical
systems and entropy

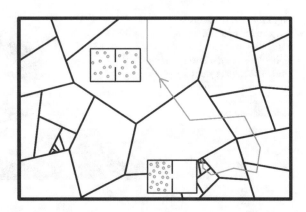

is that we have access to laws of nature which are to an important extent independent
of the particular details of implementation.

Entropy is well described in terms of the coarse graining of the phase space—as
the logarithm of the number of possible microstates that appear indistinguishable
at macroscopic level (Fig. 17.5). This allows us to speak about entropy in any possible dynamical system, as long as there is such a coarse graining. For long time I
considered that this makes thermodynamics completely reducible to the dynamical
system, but this picture also shows that thermodynamics is shared by many dynamical systems, by its generality it goes far beyond any particular set of fundamental
laws.

Thermodynamics is an example of a theory that, although a consequence of the
fundamental laws, at the same time is independent of them. We can know it without
knowing the low level laws, and as our ideas about the fundamental laws changed,
it still made sense to talk about thermodynamics without much change. And if there
are other universes with completely different fundamental laws, thermodynamics
may still be there. Similarly, classical mechanics can be seen both as emergent from
quantum mechanics, or as a stand-alone theory. To understand and describe a higher
level not only we don't need to know the underlying lower level, but the laws of the
higher level can be seen as fundamental as those of the lower level. They may be
reducible to them, or built on top of them, but they have their own logic.

These examples show that, while we can think the hierarchy of the levels of reality
to be a pyramid built on top of the most fundamental laws, these levels may float
independently, and we can interchange levels from distinct pyramids. This is what
allows us to try different theories of the microscopic level, without risking to destroy
the achievements obtained so far in mechanics, chemistry, biology etc.

17.7   Negative Knowledge

While the bottom-up approach of building the pyramid of knowledge proved to be very fruitful and able to explain many high level observations by reducing them to low level principles, sometimes this can make us wander blindly in the search of an explanation which simply can't exist.

A paradigmatic example is the search of a causal and realistic theory which was expected by many to exist, and which would produce the same results, in particular the same probabilities, as those of quantum mechanics. The idea was to find a theory relying on variables which are well-defined at each instant, and which can give the quantum probabilities in a way similar to how classical mechanics combined with statistics can explain the emergence of thermodynamics. This program was proved to fail by Bell's theorem [10], unless we give up other cherished principles. In particular, such theory can exist, but only as long as the variables are nonlocal and hidden—which means, in the context of relativity, that they are somewhat contrary to the idea of causality. The alternative seems to be to give up the very hope of having a realistic description, and admit as real only the probabilities.

Bell's theorem and other impossibility results like the Kochen-Specker(-Bell) theorem [11, 12] don't rely on a particular table of laws. They are universal, they are a priori truths. And combined with the experimental data, Bell's theorem rules out local hidden-variables theories [13].

These theorems made almost all physicists give up local hidden-variables theories and the non-contextual ones. They could give up this hope because quantum mechanics works perfectly well in practice, and to some is satisfactory enough to refrain on asking foundational questions and focus instead on the results.

Other no-go results are known to exist in classical general relativity, which was proven to lead to the occurrence of singularities [14–18]. A widespread view is that by predicting singularities, general relativity predicts its own breakdown, and one should replace it with something else, or maybe a quantum theory of gravity would remove the singularities. However, there is no clear evidence that singularities lead to the inconsistency or the breakdown of the theory. In fact, general relativity can be formulated in an invariant and meaningful form both geometrically and physically, without leading to infinities—hence the singularities, while still present, pose no insurmountable problems (see [1] and references therein).

A quite different position is taken for *quantum field theory*. Its foundations were seriously shaken by Haag's theorem, which showed that the interaction picture is inconsistent [19, 20]. Moreover, the theory leads to infinities too, proving that one should not trust the reductionism of fields to particles. So it seems that quantum field theory has problems of the same size as those of general relativity and local hidden variables in quantum mechanics combined, yet we don't rush to reject it. This is because the overwhelming empirical evidence led physicists to ignore Haag's theorem, and to discover methods of renormalization which make the infinities more digestible. Quantum field theory is an example where empirical success made us ignore a priori truths. The lowest-level of the pyramid of physics seems to be rooted in a not so perfect way in the ground of mathematics.

In religion, *apophatism* or *negative theology* is an old approach stating that we can't know what God is, we can only know what attributes God doesn't have. Science takes a similar position regarding our knowledge of the laws of nature: we can't know for sure what those laws are, we can only know how to reject them. We don't know the Tablet of the Law, and we can never be certain we know it. Similar to apophatic theology, in science we can only find out what the laws are not, through the no-go theorems and through experiment and observation, whose main role is to refute theories. Other no-go results like Gödel's incompleteness theorem and Turing's theorem on the halting problem limit the power of reductionism. All these are part of the metatheory, rather than of the tablet of the law.

17.8 Bottom Versus Top—Kōantum Mechanics

Quantum mechanics is incredibly successful in describing the lowest levels of reality accessible to our experiments and in revolutionizing our technology, and since its discovery little was added to its formulation. Yet, not everyone feels satisfied with our understanding of its foundations, particularly of some apparent paradoxes, each of which reminding us of a *kōan* (zen paradox). This part of quantum mechanics should be a subfield in its own rights, and I think it deserves the name *kōantum mechanics*.

One of the most important and puzzling features is the fact that the world appears classical to our direct experience, rather than being populated by Schrödinger cats. One should expect to see a world full of superpositions of classical objects, a world of Schrödinger cats, but instead we see a classical prosaic reality. The classical level of reality seems to defeat the quantum level.

This is paradoxical, since the Schrödinger equation, governing the quantum, seems to be violated. A quantum measurement device is a classical apparatus which extracts information from the quantum level. When the observed quantum system is in a superposition of states which the apparatus is able to distinguish, Schrödinger's equation predicts a superposition of states of the apparatus, corresponding to each of the possible states of the observed system. However, we never see such a superposition, which shows that something happens to select only one possible result. This conflict between the classical apparatus an the observed quantum system led physicists to postulate that the evolution of the quantum system breaks down during the measurement, resulting in the *wavefunction collapse*—this ensuring the detection of only one of the states the apparatus can distinguish, rather than a superposition which seems to be predicted.

It is true that if we don't ignore the quantum degrees of freedom of the apparatus, which in fact is not really a classical system, we can advance in solving this puzzle. According to the theory of *decoherence*, the quantum interactions between the environment (which includes the apparatus) and the observed system turn quantum probabilities into classical ones. Unfortunately, this only solves the problem partially, since the choice among the possible results, while being now governed by classical probabilities, still requires a collapse—to reduce the statistical ensemble obeying classical probabilities to a single outcome. Alternatively, we can consider that all possible results are realized in different worlds that appeared due to the measurement. But even in this case, each such world includes its own collapse. And this is a problem, because it is very strange to claim that we have a universal law like Schrödinger's equation, which can be broken from time to time, apparently with no understood cause.

Moreover, the wavefunction collapse has some other serious problems, in particular it leads to violations of the conservation laws [3].

The universe seems to live in a very small subset of the Hilbert space consisting of the states that look almost classical. There is no apparent reason for this. Why would the quantum level be restricted to ensure the emergence of the classical level? It looks like not only the quantum level determines the classical level, but also that the classical level, by its mere existence, determines in its turn the more fundamental quantum level (Fig. 17.6). And the classical level always seems to win at the macroscopic scale. And since humans are the ones performing the experiments, it seems that we are either able to break the Schrödinger equation, or that there is no reality beyond the quantum probabilities.

Therefore, the relation between the classical level and the quantum one is of mutual determination [21]. The most fundamental level we know, the wavefunction whose dynamics is described by the Schrödinger equation, is not directly accessible. We can only measure properties of the wavefunction, but not its complete state. The true state of the wavefunction is constrained by our choice of the properties to measure. This interdependence and the limits of observability are summarized in Fig. 17.7.

However, it is possible for the dynamics to always happen according to the Schrödinger equation, even when measurements are performed, and at the same time to obtain definite outcomes [2, 3, 22, 23]. But the only way to do this is if we see measurements as partially deciding the initial conditions of the universe, or that the classical level imposes some constraints on the quantum level, which have to apply to the entire history [24]. In other words, the measurements impose a sort of initial conditions, which apply to the entire history of the universe, but are imposed at the moment of measurement—*delayed initial conditions*.

Fig. 17.6 The interdependence of the classical and the quantum levels

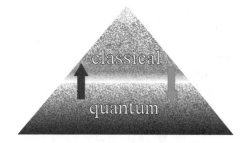

Fig. 17.7 Observations don't determine the true state, which is never directly observable. But each observation refines the set of possible wavefunctions that could give the obtained outcomes

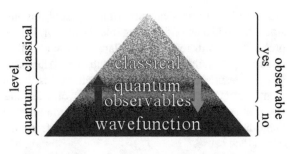

Fig. 17.8 The quantum dynamics described by the Schrödinger quation can be made compatible with the apparent collapse. Each measurement keeps only a subset of the possible states, and rules out those that don't correspond with the new outcome

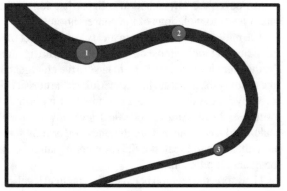

This retrocausality only happens at the unobserved level of the quantum dynamics. It doesn't change the observed past, but it refines what could be the state of the wavefunction that resulted in those observations. The wavefunction itself is not observable, only the outcomes of the measurements—the *quantum observables*. And the measurements don't determine completely the state of the wavefunction, its exact "shape", they only determine a set of wavefunction that could possibly give those outcomes. And future observations refine this information, each of them reduces the set of admissible wavefunction to a subset (Fig. 17.8). Another useful way to understand this as not being contrary to causality is to think at the universe as a *block world*, a timeless structure which includes the time dimension within, for example the relativistic spacetime. This picture allows us to think the initial conditions of the hidden wavefunction as not being fully specified at the big bang, but are events related to the measurements distributed in spacetime. The always incompletely known wavefunction has to satisfy all these conditions in a *globally consistent* way [21, 23].

Quantum mechanics is based on a wavefunction which can't be completely known, which may seem contrary to our usual view of science as being based only on testable theories. This may be a reason why many quantum theorists consider as fundamental the level of observables, and see the Schrödinger equation only as a tool to calculate the probabilities. Any attempt to find a deeper explanation is therefore called *interpretation*, even if such attempts may come with different ontologies.

But what kind of interactions count as measurements and lead to this reduction of the Hilbert space to a much smaller subset that gives the appearance of the classical world? I proposed a program to identify such interactions, along with some experiments to test the simplest one, in [25]. But even in the case we will resolve this problem, if the wavefunction is real rather than mere probability, causality as we know it has to be reconsidered [24].

17.9 Wandering Towards a Goal

A computer program is a good example of a dynamical system which is independent of its implementation on the particular lower level system on top of which it is built. You can create a computer program, compile it on a computer, and use copies of it on other computers without having to create it again. The computer's hardware may be seen as a fundamental level, and the programs running on that computer as a higher level, to which the hardware is simply irrelevant, being only a support, a physical implementation.

A computer is a machine functioning according to the physical laws. The program running on it tells it what to do, and it does, but at the same time, the physical laws tell the circuits of the hardware what to do, and they do. The fundamental physical laws governing the computer are not conflicting with the instructions of the program. For all purposes, the software controls the hardware, in a striking similarity to how the human mind controls the body. The computer follows the physical laws, and at the same time it seems to follow the goals of the program. One may say that these are not the goals of the program, but rather the goals of the programmer, or those of the employer of the programmer, or of the customers. This is also true, but couldn't it be possible for a program to just arise out of randomness, together with its goals?

It is possible that a blind dynamical system, like our universe seems to be, to contain higher level systems apparently having their own goals, and that these goals arise from the dynamics itself, without requiring a consciousness.

In an unstable world where the higher level systems are at the mercy of the low level immutable laws, no such system is eternal. It follows that the most likely goal-oriented systems are those having as goal the creation of systems similar systems to them—precisely the self-reproducing systems. Such systems are studied for long time, one of the pioneering works being von Neumann's *theory of self-reproducing machines* [26], and tremendous progress has been made since them. This seems to be the attribute of every living being, while most of the other goals it may have serve to the purpose of survival and reproduction. And even if self-reproduction is not perfect, leading to mutations, this imperfection is magnificently exploited to obtain more complex self-replicating systems. The *theory of evolution* is based on the fact that the mutations leading to behaviors more adapted to the environment are most likely to survive and replicate.

Some objections that are usually brought against even the most recent forms of Darwin's theory of evolution include the practical impossibility of giving a low level description of organisms and the process of evolution, the apparent impossibility to falsify things that already happened on very long periods of time, and the apparent improbability that blind chance can explain it [27]. I think that it is not necessary to give a low level description and to travel back in time to see that it really happened like this, it is enough to prove its likeliness in principle, and to show that our universe satisfies the conditions in the hypothesis of the proof. But still, one may say that it is very improbable for such system to appear simply out of blind chance.

A brilliant attempt to prove Darwin mathematically is due to Chaitin, who developed a model based on computer programs subject to random mutations. His result is amazing: exhaustive search requires an exponential time $\approx 2^N$, while cumulative evolution a time between $\approx N^2$ and $\approx N^3$, very close to the time required by choosing the next mutation by intelligent design, which is $\approx N$ [28]. But while I think Chaitin's idea is really smart, I am not entirely satisfied with his model, because in finding the mutations and selecting the fittest program, it relies on *oracles*, which would require too much computation if implemented as programs, so they seem just a placeholder for divine intervention or intelligent design.

If the universe is a dynamical system, to find out the probability it contains life one should find out the states containing life, and count the possible histories leading to these states. If the laws are deterministic, then the possibility of life has to be already encoded in the initial conditions. If the laws are non-deterministic, involving random jumps, then the initial conditions combined with the random jumps must include the complete information regarding the future emergence of life. In both cases, it seems very unlikely and circular—all living beings that exist, existed, or will exist at some time, must already exist encoded in the initial conditions combined with the updates of initial conditions included in the jumps. It is understandable that some find more likely that the initial conditions are fine tuned or designed, and that further jumps actually are divine interventions with the purpose of leading to life as we know it. But is this necessarily the case?

What is certain is that there are histories of the universe— trajectories in the phase space—which at some points contain life, and even human-like intelligent beings. After all we would are here. No matter how small is the set of trajectories containing life, it is definitely not an empty set.

Then, one solution is to assume that all possible trajectories in the phase space represent histories of different universes, which are not only possible, but actual, and we can only live in such a history containing beings like us.

One may think that the hypothesis that all possible states of the universe are realized somewhere as exceedingly uneconomical. In fact one needs much more resources to define a point in the phase space, than to define the entire phase space— even if the phase space or state space is infinite-dimensional, its definition may fit on the tablet of the law. Another possibility is that the universe is very vast, perhaps infinite. Then we can see it as being made of an infinite number of dynamical systems combined together, and by this, from point to point, the lucky trajectories containing life are realized. In particular, considering the huge number of planets in the known universe, it is not so surprising that some of them are favorable to life. So it is also understandable that for others this hypothesis is more economical than postulating a God to explain the occurrence of life and consciousness.

Evolution by random mutation and selection is another example of processes that can emerge in a large class of dynamical systems resembling our universe—just like entropy, or like how a computer program can work on different types of computers.

17.10   Sentient Observers

The most directly accessible phenomenon to our mind, yet one of the least understood, is our consciousness—in fact, our entire experience takes place in our consciousness. Even what we call objective reality, is different for each of us. The only reason we have the feeling that we all share the same world is because there are persistent phenomena which we experience together with the others, about which we are able to discuss because we developed and shared various conventions. But even when we discuss about the simplest objects in nature, we actually discuss about our internal, subjective representations. At the bottom of our understanding doesn't lie reality, but consciousness, subjectivity, even when we talk about about things we call objective. Even if everything is built on the top of a real word, our representations are subjective, and what we call objectivity is just consensus between different conscious subjects. In spite of the ubiquity of consciousness in our representations of reality, consciousness is one of the greatest mysteries, and by far the most important for humans.

There is something about consciousness that some, including myself, find most mysterious. Is it about intelligence, imagination, creativity, or other features we usually assume are the prerogatives of humans? While very complex and important, it is conceivable that all these processes are eventually completely reducible to neural activity, and we already know that they have neural correlates— that is, they are mirrored by physical, chemical, and biological processes taking place in the brain. Is it the qualia—the unique way each of us experiences the qualities of colors, sounds, feelings? These differences can be due to the unique way the brain of each individual is wired. Is it the impression that there is a permanent center of our thoughts? This may very well be an illusion, during our lives we change so much that sometimes we are unable to recognize ourselves. And if our brain hemispheres are separated surgically, observations revealed that consciousness is split too. This identity is not permanent, is always changing, although it feels invariant, and it may be an illusion too, a trick by which Mother Nature convinces us to take care of ourselves. Is self-awareness then the essence of consciousness? If self-awareness can be reduced to having a representation of ourselves, which helps us monitor how close we are to attaining one goal or another or to endanger ourselves, then a refrigerator is self-aware, by having a representation of its temperature and acting towards maintaining it. Are we less conscious when we are aware about external objects, ignoring ourselves? If someone would have no possibility to act on the world, and will only be able to watch others' actions, would this mean she is less conscious? Is consciousness just matter organized in a very complex way, which we will be able to completely decode in time? Can more complex patterns of information hold the answer, as in the *integrated information theory of consciousness* [29, 30]?

Many features of consciousness are understood in terms of physical processes taking place in the brain, or of neural activity. It is plausible that eventually nearly all of them will be understood like this, so Chalmers calls them the *easy problems* of consciousness, reserving the term *hard problem* [31] for the most irreducible part of our subjective experiences—the sense of a self experiencing all these.

I have the same feeling, that what is most mysterious and irreducible about consciousness is the subjective experience itself, or *sentience*. One can imagine a machine talking and behaving just like us, even displaying feelings like us, yet all these being just external, with no subjective experience really taking place inside that machine—a *philosophical zombie*.

I think the main mystery is indeed the hard problem of consciousness. Many seem to understand what the hard problem is, while others seem to see no problem at all, and believe that once we will understand in detail how the mind works, the hard problem will turn out to be just an illusion, or the question itself to be meaningless. And on top of this problem is another one: there is no rigorous and objective definition of subjective experience. Perhaps whenever we try to define it in objective terms, we make it trivial and irrelevant. Science is by definition objective—all definitions have to be objective, all inferences are objective, the experiments have to be reproducible by anyone who follows the specifications, the logical or mathematical proofs have to be verifiable by other researchers. All *easy problems* of consciousness, those pertaining to functionality, which we can assume we will someday be able to monitor at the level of neurons, if not of atoms, fall within the objective scope of science. But the very notion of subjective experience seems to escape any objective definition.

Is this impossibility to give an objective definition of subjectivity a proof that the hard problem actually is a nonproblem? Is the very existence of the hard problem a matter of personal belief, which can and will eventually be deemed as nonscientific? Or rather we need a science of the subjective itself? A subjective science can't be objective—even personal experiences that admit descriptions will be reproduced by others subjectively, and there will be no assurance that we are talking about the same experiences. To some extent, this is true even for the objective science, since even objective experiences are interpreted inside the consciousness of each of us. I think the only framework in which one can define the hard problem is subjective, outside the realm of objective science.

Let us take the position that subjective experience emerges from the organization of matter, or as a property of information, like integration, and see where it leads us. We know that matter is always structured and always processes information, even if not in a way that would mean something to us. Obviously something as complex as the human mind has to be associated with a complex structure like the human brain. But what is the simplest structure or entity endowed with sentience? Does this structure need to deal with memory, creativity, representation of itself or of other structures, etc, to explain subjective experience? Does it need to be associated with a biological support?

But what is experience? Objectively speaking, the experience of a system is whatever happens to that system, its changes, its processes. But subjectively speaking, it can't be any of these, unless somehow sentience is attached to the system. But what does this even mean? Is there a type of particle or fundamental structure, physical or informational, which is the building block of consciousness, especially of sentience? Or is sentience just a ubiquitous feature of any kind of structure that processes information?

If subjective experience is separated from the material or informational structures, as in Descartes's dualism, then the other world of consciousness that shadows ours still has laws. If so, how can we prove dualism, given that the physical world doesn't seem to be influenced from the outside? Even if by some completely not yet known process we are able to detach ourselves from the physical world and have immaterial experiences, when we return our brain will "download" the information about these experiences and it will be impossible to say that they took place outside the physical world, or were just created by our brain by natural processes that happen instead of what we think is the "download".

Whether or not sentience is irreducible, it still must be associated to structure and information in an inseparable way, so that the physical correlates perfectly mirror the corresponding aspects of conscious experience. We arrive at a sort of panpsychism, with sentience intrinsically associated to the physical substance, whether it is just a property of physical objects or not. And there is no objective way to tell whether sentience is reducible or not to physics.

But what is the physical structure to which sentience is attached? Is it attached to matter, to spacetime, to quantum fields, to the wavefunction, to a particular type of structure in which matter is organized, or to everything?

Maybe sentience is attached to goals, or at least we are more aware of it in the presence of goals. We have a wide diversity of goals, most of them not being directly concerned with survival or reproduction, but maybe they were initially means to thes basic goals, and in time they became standalone goals. Various subjective experiences can be attached to attaining, anticipating, fear of missing or losing a goal. Of course, the term "goal" may lead us to the idea of intention and agency, which may suppose already consciousness. But here by "goal" I mean the objective of a task of an automaton, being it designed by man or resulted from natural evolution.

A possible way to test subjectively whether sentience is attached to goals is by mindfulness or meditation. If in meditation you change the goal from attaining your goals to simply being and observing with no attachment, then the emotions related to attaining or failing to attain it are short-circuited. You can have an ecstatic state of consciousness with no apparent relation to your goals, just from attaining the goal of not having goals that you can't control, or having as goal just observing the natural flow of events.

But our spectrum of experiences is far richer than those we can consciously identify as being related to goals, for instance the qualia of colors, sounds, feelings induced by words and memories, all these are subjective experiences which can hardly be tracked back to goals. What if each cell of our bodies has its own experiences related to its own "minigoals", and our larger experiences emerge by their integration?

If sentience is either reducible to physical structure and information, or if it is irreducible, but attached to these, then one should expect it to be present in primitive forms at each level of reality, since structure and information are also present there.

Even a photon or an electron can be said to be associated to information processing, and if there are structures that have goals, we can say even that particles have a goal—the goal of propagating according to the physical laws. This goal is always to propagate an infinitesimal step, and it is always attained in an infinitesimal time, in accord to the equations of motion. If sentience is to be associated to this goal, then would the corresponding experience be the pure bliss of always attaining it instantly?

Can subjective experiments like meditation allow us to verify the kind of structure or the level where subjective experience emerges, or if sentience is reducible or not to physical structure or information?

17.11 Why Is There Something Rather Than Nothing?

While this question is often considered meaningless or rhetorical, I think it is important and fundamental, and deserves serious consideration.

A possible answer is that by "nothing" one should understand vacuum, the universe being just a fluctuation of the vacuum [32], or the string landscape [33, 34]. But let's see if we can make sense of "nothing" without redefining it to something.

To answer this question, let's try to see what constituents or features of the universe could very well not exist. We can imagine that a universe identical to ours, but with different initial conditions, hence different history, is possible. Its history is just another path in the phase space. Even a universe in which the fundamental particles or laws are different seems equally possible. If our universe exists, there is no reason to assume that another universe, completely different from ours, can't exist too. It seems that no object, structure, or physical law from our universe, is bound to exist, or to exist in the form we know.

But there are things that exist with necessity. Consider Euclidean geometry, which exists in its abstract, mathematical form in our universe, despite the fact that space-time itself is not Euclidean. Euclidean geometry exists because we can imagine it consistently, not because it is realized by the physical laws. It can exist in any possible universe containing substructures that can imagine it. I am talking of course about mathematical existence, which is usually considered different from physical existence. Mathematical existence simply means logical consistency. When we say "between any two points on a line there is a third point", here "there is" refers to mathematical existence—in the sense that the existence of that point is consistent with the existence of the first two points, not that there is an actual physical realization of any of the three points or of the line. The usual view is that only physical existence is real, while mathematical existence is imaginary. However, mathematical objects exist, in mathematical sense, while physical existence, known to us only by empirical evidence, is unnecessary, in the sense that another universe is possible, in which that particular physical structure or law doesn't exist physically. Moreover, if consciousness is just structured information, wouldn't such a substructure contained in a mathematical structure validate the mathematical structure as existent?

Is mathematical existence the only necessary existence? Physical universes don't seem to exist with necessity, so it is legitimate to ask why they exist. But mathematical structures exist with necessity, in mathematical sense, and they are a priori truths.

If mathematical structures exist anyway, and if the universe is isomorphic to a mathematical structure [9, 21], do we need something more to explain why there is something rather than nothing? This leads us straight to Tegmark's *mathematical universe hypothesis* [35–37], which posits that physical existence equals mathematical existence —in other words, logical possibility equals reality. Accordingly, we are just substructures of such a mathematical structure, we observe the structure as it appears to us, and ask questions like the one in the title.

Alternatively, one can say that all there is is digital information [7, 28, 38–42]. I find the idea of mathematical structures more appealing than digital philosophy, because information seems to require decoding, translation into a meaning, while mathematical structures just are.

We all know that in order to run a computer program, one needs a computer. But both the program and the computer are just dynamical systems. The hardware itself changes from one state to another, just like a program. What is the support on which the hardware runs? The universe? Then what is the support on which the universe runs? We have to stop somewhere, and I think the only way to stop is to admit that the universe is a dynamical system which runs by itself. It is pure information processing, pure mathematical deduction, pure computation.

Information is finite, while mathematical structures, even if they can be seen as processing information, can encode unlimited information. Attempts to reduce to digital information the mathematical structure describing the universe, or actually being the universe, seems to be at odds with what we know about the limits of computations and with Gödel's theorem. They may appear as encoding and processing information, and for a subsystem, the interaction with other subsystems encodes information. But mathematical structures are not limited to digital information, which is a good thing, since our most successful theories of physics, quantum mechanics, general relativity, and the Standard Model, all seem to require the continuum, or at least we don't know yet how to reduce any of them to a discrete or digital theory. This doesn't exclude the possibility that the underlying mathematical structure of the universe is discrete or even finite, or that it is pure computation as in the digital philosophy of the above cited authors. But I would avoid making such limiting assumptions yet.

One can object to the identification of physical existence with mathematical existence by claiming that the latter is imaginary. But I think this is a different kind of imaginary, since it is consistent. If one would dream his entire life, and the dream would be consistent, then wouldn't that person experience imaginary as real?

The question "why is there something rather than nothing?" can receive thus the following answer:

"because there are structures that can't not exist—mathematical structures".

17.12   What Breathes Fire into the Equations?

Few are satisfied with proposals like Tegmark's that mathematical existence is enough. Hawking asks [43]

> Even if there is only one possible unified theory, it is just a set of rules and equations. What is
> it that breathes fire into the equations and makes a universe for them to describe? The usual
> approach of science of constructing a mathematical model cannot answer the questions of
> why there should be a universe for the model to describe. Why does the universe go to all
> the bother of existing?

Perhaps the main reason why we find the idea of a mathematical universe uneasy is that we consider mathematics a creation of our minds, even a fantasy. Because we can imagine it, it seems that in order to be true, it has to be realized physically, we think that anything can only exist if it is made of matter. But what is matter made of? Quantum mechanics tells us that either there is no matter at all, or if there is, it is completely different from what we used to think it is. The quantum metatheorems by Bell and Kochen-Specker force us to choose between the complete nonexistence of matter or any kind of reality, and the existence of a reality which depends on distant places or on future choices. And while most scientists seem to agree that materialism won, this victory came with the price of redefining the very notion of matter, by replacing it even with its very absence: quantum mechanics says that what we used to call matter is either pure probabilities and information, or some stuff whose state depends on what measurements are made far away or in the future.

I think what worries us about the idea of a mathematical universe is in fact consistency, but mathematical structures, unlike dreams and fantasies, are logically consistent.

In addition to this, one may feel that physics and mathematics are not enough to build consciousness. But almost all features of consciousness are conceivably reducible to information processing of one sort or another. If something resists, this is subjective experience.

In the absence of an absolute ground to rely on, I think what we really know is that we are, and that there are mathematical truths. And whatever reality is, we know that it is like a mathematical structure with its equations, which has the ability of experiencing itself. Are there other ingredients needed?

17.13   The Tablet of the Metalaw

There is a fundamental level of reality, but there are also higher levels, each with its own life, and not so rooted in the lower levels and reducible to them as one may want to think. At the top of the pyramid are life and consciousness, and they should be the center of science too. A bottom-up approach may never lead to the understanding of the higher levels, and there are also top-down constraints. While it is desirable that eventually natural sciences will be founded on fundamental physics, and to apply

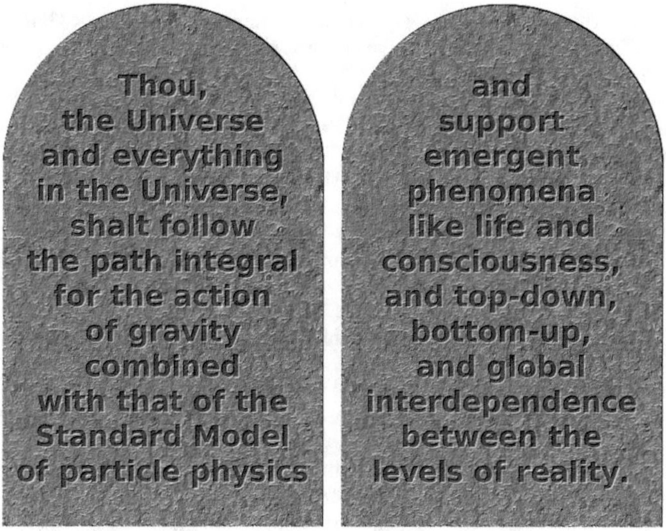

Fig. 17.9 The tablet of the law and the tablet of the metalaw

reductionism whenever possible, each science has its own empirical domain, from which it draws its laws, and against which it is tested. One should take into account the explanatory power and utility of each hypothesis within its own field at its own level of reality. In addition, one should not forget that natural sciences rely on testing the hypotheses through observation and experiment, but the source of the hypotheses is not and should never be regulated. Any activity that helps you come up with scientific hypotheses is useful, from philosophy to arts and even to myths. Hypotheses have the first word, but logical consistency and experiments have the last word.

We saw that, despite the tremendous success of science to find the tablet of the physical law and to provide a reductionist view of other disciplines to fundamental physics, there is more to be said. I think it is justified to consider a second tablet, the *tablet of the metalaw* (Fig. 17.9). This tablet includes emergence, metatheorems, the relative interdependence and independence of various levels of reality. It takes into account both the bottom-up and the top-down constraints. It may even include a subjective science of the subjective experience.

References

1. Stoica, O.C.: Singular general relativity. Ph.D. thesis, Minkowski Institute Press, (2013). arXiv:math.DG/1301.2231
2. Stoica, O.C.: On the wavefunction collapse. Quanta. **5**(1), 19–33 (2016). https://doi.org/10.12743/quanta.v5i1.40
3. Stoica, O.C.: The universe remembers no wavefunction collapse. Quantum Stud. Math. Found. (2017). arXiv:1607.02076

4. Stoica, O.C.: The Standard Model Algebra (2017). Preprint arXiv:1702.04336F
5. Bates, S., Weinstein, A.: Lectures on the Geometry of Quantization, vol. 8. American Mathematical Society, Rhode Island (1997)
6. Arnowitt, R., Deser, S., Misner, C.W.: The dynamics of general relativity. In: Gravitation: An Introduction to Current Research, pp. 227–264. Wiley, New York (1962)
7. Wolfram, S.: A new kind of science, vol. 5. Wolfram Media Inc., Champaign (2002)
8. Stoica, O.C.: World theory. Phil. Sci. Arch. (2008). philsci-archive:00004355
9. Stoica, O.C.: And the math will set you free. In: Trick or Truth? The Mysterious Connection Between Physics and Mathematics, pp. 233–247. Springer, Berlin (2016). arXiv:1311.0765
10. Bell, J.S.: On the Einstein-Podolsky-Rosen paradox. Physics **1**(3), 195–200 (1964)
11. Bell, J.S.: On the problem of hidden variables in quantum mechanics. Rev. Mod. Phys. **38**(3), 447–452 (1966)
12. Kochen, S., Specker, E.P.: The problem of hidden variables in quantum mechanics. J. Math. Mech. **17**, 59–87 (1967)
13. Aspect, A.: Bell's inequality test: more ideal than ever (1999)
14. Penrose, R.: Gravitational collapse and space-time singularities. Phys. Rev. Lett. **14**(3), 57–59 (1965)
15. Hawking, S.W., Penrose, R.W.: The singularities of gravitational collapse and cosmology. Proc. Roy. Soc. London Ser. A **314**(1519), 529–548 (1970)
16. Hawking, S.W.: The occurrence of singularities in cosmology. Proc. Roy. Soc. A-Math. Phy. **294**(1439), 511–521 (1966)
17. Hawking, S.W.: The occurrence of singularities in cosmology II. Proc. Roy. Soc. A-Math. Phy. **295**(1443), 490–493 (1966)
18. Hawking, S.W.: The occurrence of singularities in cosmology III. Causality and singularities. Proc. Roy. Soc. A-Math. Phy. **300**(1461), 187–201 (1967)
19. Haag, R: On quantum field theories. Kgl. Danske Videnskab. Selakab, Mat.-Fys. Medd. 29 (1955)
20. Hall, D., Wightman, A.S.: A theorem on invariant analytic functions with applications to relativistic quantum field theory. I kommission hos Munksgaard (1957)
21. Stoica, O.C.: The Tao of it and bit. In: It From Bit or Bit From It?: On Physics and Information, pp. 51–64. Springer, Berlin (2015). arXiv:1311.0765
22. Stoica, O.C.: Flowing with a Frozen River. Foundational Questions Institute, "The Nature of Time" essay contest. http://fqxi.org/community/forum/topic/322 (2008). Last accessed 27 Jan 2018
23. Stoica, O.C.: Global and local aspects of causality in quantum mechanics. In: EPJ Web of Conferences, TM 2012—The Time Machine Factory [unspeakable, speakable] on Time Travel in Turin, vol. 58, p. 01017. EPJ Web of Conferences, Sept 2013
24. Stoica, O.C.: Quantum measurement and initial conditions. Int. J. Theor. Phys. 1–15 (2015). arXiv:quant-ph/1212.2601
25. Stoica, O.C.: Searching for microscopic classical cats, Preprint arXiv:1604.05063 (2016)
26. Von Neumann, J., Burks, A.W., et al.: Theory of self-reproducing automata. IEEE Trans. Neural Netw. **5**(1), 3–14 (1966)
27. Moorhead, P.S.: Mathematical Challenges to the Neo-darwinian Interpretation Of Evolution. A symposium held at the Wistar Institute of Anatomy and Biology, 25–26 April, 1966. (1967)
28. Chaitin, G.: Proving Darwin: Making Biology Mathematical. Vintage, New York (2012)
29. Tononi, G., Boly, M., Massimini, M., Koch, Ch.: Integrated information theory: from consciousness to its physical substrate. Nat. Rev. Neurosci. **17**(7), 450–461 (2016)
30. Tegmark, M.: Consciousness as a state of matter. Chaos Solitons Fractals **76**, 238–270 (2015)
31. Chalmers, D.J.: Facing up to the problem of consciousness. J. Conscious. Stud. **2**(3), 200–219 (1995)
32. Krauss, L.M.: A Universe From Nothing: Why There is Something Rather Than Nothing. Simon and Schuster, New York (2012)
33. Susskind, L.: The Cosmic Landscape: String Theory and the Illusion of Intelligent Design. Back Bay Books, New York (2008)

34. Hawking, S., Mlodinow, L.: The Grand Design: New Answers to the Ultimate Question of Life. Bantam Books, New York (2010)
35. Tegmark, M.: The mathematical universe. Found. Phys. **38**(2), 101–150 (2008)
36. Tegmark, M.: Is the theory of everything merely the ultimate ensemble theory? Ann. Phys. **270**(1), 1–51 (1998)
37. Tegmark, M.: Our Mathematical Universe: My Quest for the Ultimate Nature of Reality. Knopf Doubleday Publishing Group, New York (2014)
38. Wheeler, J.A.: Law Without Law. In: Wheeler, J.A., Zurek, W.H. (eds.) Quantum Theory and Measurement, pp. 182–213. Princeton University Press, NJ (1983)
39. Wheeler, J.A.: On recognizing 'law without law, Oersted Medal response at the joint APS-AAPT meeting, New York, 25 January 1983. Am. J. Phys. **51**, 398 (1983)
40. Wheeler, J.A.: Information, physics, quantum: the search for links. In: Complexity, Entropy and the Physics of Information. The Proceedings Of The 1988 Workshop On Complexity, Entropy, And The Physics Of Information, vol. 8. Santa Fe, New Mexico, 29 May–10 June 1989. Westview Press, Colorado (1990)
41. Fredkin, F.: An introduction to digital philosophy. Int. J. Theor. Phys. **42**(2), 189–247 (2003)
42. Rucker, R.: Mind Tools: The Five Levels Of Mathematical Reality. Courier Corporation, New York (2013)
43. Hawking, S.W.: A Brief History of Time. (1988)

Chapter 18
Wandering Towards a Goal: The Key Role of Biomolecules

George F. R. Ellis and Jonathan Kopel

18.1 Physics Versus Biology

How can a universe that is ruled by natural laws give rise to aims and intentions? Whether or not a human observer exists, the natural laws would continue to operate as they are indefinitely. The key difference between physics and biology is function or purpose. There is, in the standard scientific interpretation, no purpose in the existence of the Moon[1] or an electron or in a collision of two gas particles. By contrast, there is purpose and function in all life [21]:

> Although living systems obey the laws of physics and chemistry, the notion of function or purpose differentiates biology from other natural sciences. Organisms exist to reproduce, whereas, outside religious belief, rocks and stars have no purpose. Selection for function has produced the living cell, with a unique set of properties that distinguish it from inanimate systems of interacting molecules. Cells exist far from thermal equilibrium by harvesting energy from their environment. They are composed of thousands of different types of molecule. They contain information for their survival and reproduction, in the form of their DNA.

How does purpose or function emerge from physics? At the macro level, in higher animals and human beings, via adaptive neural networks [25] and physiological systems [38]. At the micro level, through epigenetic effects in cell development [17] via gene regulatory networks [17] and through adaptive effects in signal transduction networks [24] and synapses [25]. Although many have argued that these are all based only on the lower levels through interactions of specific molecules, particularly

G. F. R. Ellis (✉)
Mathematics Department, University of Cape Town, Cape Town, South Africa
e-mail: gfrellis@gmail.com

J. Kopel
Texas Tech University Health Sciences Center (TTUHSC), Lubbock, USA

[1]It is true that the existence of the Moon was probably essential for the origin of life as we know it, so one might claim that the purpose of the Moon was to enable life on Earth to emerge from the sea to dry land through its effects on tides. However the Moon is unaware of this effect: it was, as far as the Moon was concerned, an unintended by product of its orbital motion round the Earth.

© Springer International Publishing AG, part of Springer Nature 2018   227
A. Aguirre et al. (eds.), *Wandering Towards a Goal*, The Frontiers Collection,
https://doi.org/10.1007/978-3-319-75726-1_18

proteins [36] and nucleic acids [51] , in fact biological systems are as much shaped by contextual effects as by bottom-up emergence [32]. These molecules are key, but they get their effectiveness because of the context in which they exist. As Denis Noble argued,

> Genes, as DNA sequences, do not of course form selves in any ordinary sense. The DNA molecule on its own does absolutely nothing since it reacts biochemically only to triggering signals. It cannot even initiate its own transcription or replication. It cannot therefore be characterised as selfish in any plausible sense of the word. If we extract DNA and put it in a Petri dish with nutrients, it will do nothing. The cell from which we extracted it would, however, continue to function until it needs to make more proteins, just as red cells function for a hundred days or more without a nucleus. It would therefore be more correct to say that genes are not active causes; they are, rather, caused to give their information by and to the system that activates them. The only kind of causation that can be attributed to them is passive, much in the way a computer program reads and uses databases? [30].

To be clear and concise, this paper will focus on voltage gated ion channels. We will first look at the difference between the logic of physics and biology, then at the molecules that make this difference possible, and finally at the way such molecules have come into being and allowed physical processes to generate biological activity. We take for granted that living systems are open non-equilibrium systems. That alone does not characterise life: famously, even a burning candle satisfies those conditions. Something more is required.

18.2 Logic of Physics

Physical laws determine evolution of a physical system in a purposeless inevitable way. Let the relevant variables be X and the evolution dynamics be given by $H(C, X, t)$ where the context C is set by initial and boundary conditions, then that dynamical law determines later states from earlier states:

If at time t_1, $X = X(t_1)$, then at time t_2, $X = H(C, X(t_1), t_2)$.   (18.1)

They may often be represented by suitable phase planes. Here are some examples:

- Example 1: The dynamics of classical systems such as a pendulum [3].
- Example 2: The dynamics of celestial objects governed by gravity [8, 45].
- Example 3: Gases and kinetic theory: gas molecule motion leads to predictable macroscopic behaviour [35].
- Example 4: Electron flows in transistors [42] and computers [46].

The key point is that in physics there are fixed interactions that cannot be altered, although we can to some extent decide what they act on and so what their outcome will be (as in the case of the pendulum and the computer). In the end, daily life is governed by Newton's laws of motion and Galileo's equations for a falling body, together with Maxwell's equations[2]:

$$\nabla \cdot \mathbf{E} = 4\pi\rho, \qquad \nabla \times \mathbf{E} = -\frac{1}{c}\frac{\partial \mathbf{B}}{\partial t}, \tag{18.2}$$

$$\nabla \cdot \mathbf{B} = 0, \quad \nabla \times \mathbf{B} = \frac{1}{c}\left(4\pi\mathbf{J} + \frac{\partial \mathbf{E}}{\partial t}\right) \tag{18.3}$$

relating the electric field $\mathbf{E}$, magnetic field $\mathbf{B}$, charge ρ, and current $\mathbf{J}$, and nothing can change those interactions. The equation of motion for a particle with charge e and mass m, and velocity $\mathbf{v}$ due to the electromagnetic field $\mathbf{E}$, $\mathbf{B}$ and gravitational field $\mathbf{g}$ is given by

$$\mathbf{F} = m\frac{d\mathbf{v}}{dt} = e\{\mathbf{E} + \mathbf{v} \times \mathbf{B}\} + m\mathbf{g}. \tag{18.4}$$

Equation (18.1) represents the solutions that necessarily follow from (18.2–18.4), proceeding purposelessly on the basis of the context C and initial data $\mathbf{X}(t_1)$.

18.3 Logic of Life

Life collects and analyses information in order to use it to plan and execute purposeful actions in the light of memory (stored information) [12, 21]. This is true from amoeba [2] to all animals [12] to humans [25]. Even plants do something like this, for example in the case of heliotropism (tracking the sun's motion across the sky) through cell elongation due to the phytohormone auxin [41].

18.3.1 Information Usage

Information use is based on contextually informed logical choices of the form

$$\text{Given context } C, \quad \text{IF } T(\mathbf{X}) \text{ THEN } F1(\mathbf{Y}), \quad \text{ELSE } F2(\mathbf{Z}), \tag{18.5}$$

where $\mathbf{X}$, $\mathbf{Y}$, and $\mathbf{Z}$ may be the same or different variables and $T(\mathbf{X})$, $F1(\mathbf{Y})$ and $F2(\mathbf{Z})$ are arbitrary functions that can include any logical operations (AND, OR, NOT, NOR, and so on) and mathematical operations. Thus they might be "If the cat is in, turn the light out, else call the cat in" or "If the calculated future temperature T will be too high and there is no automatic control system, then reduce the fuel flow F manually" (a default unstated "ELSE" is always to leave the status quo). In the case of flowering plants it might be "if the sun is shining, open; if not, close".

---

[2]I am avoiding discussing the extra complications introduced by quantum mechanics at this point. This will be important below.

The key point is that the functions $T(\mathbf{X})$, $F1(\mathbf{Y})$ and $F2(\mathbf{Z})$ are not determined by the underlying physical laws; they can be shaped jointly by evolutionary and developmental processes [16, 17] to give highly complex outcomes (ranging from phenotype-genotype maps [50] to the citric acid cycle [2, 26] to physiological systems [12, 38]), or can be planned by human thought to produce any desired outcome [11, 20]. Unlike the case of physical laws, where the relevant interactions cannot be changed or chosen because they are given by Nature and are invariable, these interactions can fulfil widely varying biological or social or mental purposes. It is their arbitrary nature, essentially similar to Turing's discovery that a digital computer can carry out arbitrary computations, that allows this flexibility.

Biological examples of logical processes are

- Chemotaxis: bacteria detect gradients and move away from poisons and towards nutrition [1].
- Bee dances: If food is found, bees signal its location to the rest of the hive by a waggle dance [39]. No dance implies no food has been found.
- epigenetics: gene expression at lower levels is controlled at lower levels by gene regulatory networks to meet higher level needs [17, 31, 32] (as illustrated in Figs. 18.1 and 18.2).
- Human planning of future actions on the basis of expected outcomes based on logical choices [20], for example computer aided design and manufacture of an aircraft on the basis of chosen design criteria [15].

18.3.2   Phase Transitions

A physicist might suggest that what is proposed here is in fact a part of physics through the idea of phase transitions, such as solid/liquid/gas transitions for a substance S [9]. These generically have a form like

$$\text{GIVEN pressure } P \text{ and temperature } T, \text{ IF } \{P, T\} \in S_{P,V} \text{ THEN S is solid,}$$
$$\text{ELSE IF } \{P, T\} \in L_{P,V} \text{ THEN S is liquid, ELSE S is gaseous. } (18.6)$$

Here the context is represented by the pressure P and temperature T, and $S_{P,V}$, $L_{P,V}$ and $G_{P,V}$ are subsets of the (P, V) plane. At first glance this looks like it has the form (18.5). However there is a crucial difference. The regions $S_{P,V}$, $L_{P,V}$ and $G_{P,V}$ are completely fixed by the physics of the substance S independent of the environment, and cannot be altered, whereas in (18.5), $T(\mathbf{X})$, $F1(\mathbf{Y})$ and $F2(\mathbf{Z})$ are contextually set and in some way adapted to the environment. As shown by the above examples, they have a flexibility that is completely missing in (18.6). This is physical rather than logical determination.

18.3.3 Physical Realisation

The hierarchy of structure that underlies existence of life [15] is indicated in Fig. 18.1. The kind of branching logic indicated in (18.5) occurs at each level in this hierarchy. It occurs at the lower levels via mechanisms such as voltage and ligand gated ion channels, molecular recognition via lock-and-key mechanisms, and synaptic thresholds, as discussed below.

When built into gene regulatory networks, signal transduction networks, metabolic networks, and neural networks, logic gates realised in one of these ways at the lower levels then lead to higher order logical operations such as occur in epigenetic circuits and the functioning of the brain. Within these circuits exists a unique interaction of constraints and interactions between higher and lower levels within any given system. As Denis Noble described,

> Where do the restraints come from in biological systems? Clearly, the immediate environment of the system is one source of restraint. Proteins are restrained by the cellular architecture (where they are found in or between the membrane and filament systems), cells are restrained by the tissues and organs they find themselves in (by the structure of the tissues and organs and by the intercellular signalling) and all levels are restrained by the external environment. Even these restraints though would not exhaust the list. Organisms are also a product of their evolutionary history, i.e. the interactions with past environments. These restraints are stored in two forms of inheritance—DNA and cellular. The DNA sequences restrict which amino

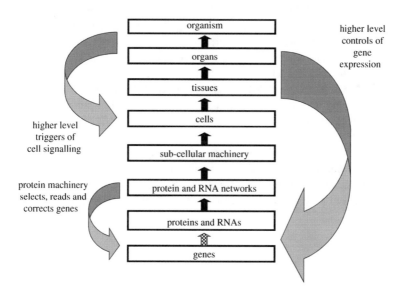

Fig. 18.1 Epigenetic and physiological control of lower level processes by higher level contexts, which control which logic will operate at lower levels by switching genes on and off. From [31] (with permission)

acid sequences can be present in proteins, while the inherited cellular architecture restricts their locations, movements and reactions [30]

Figure 18.2 shows how logical operations of a lower level module in a metabolic pathway can be regulated by higher order circuits through transcription factors that initiate and regulate the transcription of genes. They can be "on" (able to bind to DNA) or "off", thereby controlling the transcription of genetic information from DNA to messenger RNA. The local transcription factor TF_2 is sensitive to an intermediate metabolite X_n and modulates the synthesis of enzymes (powerful biological catalysts that control the rate of reactions) in the pathway. Thus they embody logic of the form

$$IF\ TF_2\ on,\ THEN\ E_1 \rightarrow X_3,\ ELSE\ NOT \qquad (18.7)$$

which is logic of the form (18.5) operating locally inside the module. However the global regulator TF_1, sensitive to higher level variables such as heart rate or blood pressure, can modulate (i) the synthesis of intermediate enzymes, (ii) the synthesis of the local transcription factor TF_2, or (iii) both. Thus overall the internal logic of the module acts to produce a "black box" whereby conversion of X_1 to X_n is controlled by TF_1:

$$IF\ TF_1\ on,\ THEN\ X_1 \rightarrow X_n,\ ELSE\ NOT \qquad (18.8)$$

This is again a relation of the form (18.5), but at a higher level (TF_1 is a higher level variable). Thus lower level logic circuits such as (18.7) can be used to build up higher level logic such as (18.8). This is the process of *abstraction* in a modular hierarchy [10], whereby internal workings are hidden from the external view (TF_2, E_2, X_2, and X_3 are local variables whose values are not known to the external world and do not occur in (18.8)). All that matters from the system view is the logic (18.8) whereby $TF1$ controls conversion of X_1 to X_2. This process of *black boxing* ([4]:Sect. 6) of lower level logic to produce higher level logic in biology contrasts strongly with the process of *coarse graining* [35] to produce higher level variables and effective laws out of lower level variables and effective laws in physics.

This kind of regulation of lower levels due to higher level conditions can occur between any adjacent level in the hierarchy of structure and function.[3] All these logical operations are based in physics, but are quite different than the logic of physical laws *per se*. How are they realised through the underlying physical stratum?

---

[3]This is like the way there is a tower of virtual machines in a digital computer, with a different formal logic operational at each level [15].

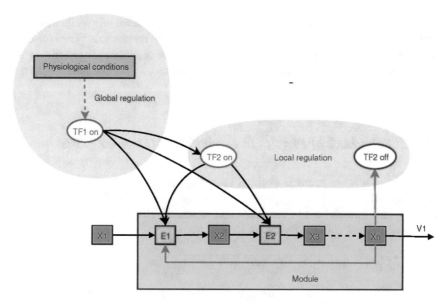

Fig. 18.2 Global and local regulation of a metabolic pathway. Logical structure of a simple metabolic pathway module converting food to energy and protein building blocks, showing the many proteins involved, as well as the top-down effect of the large-scale physiological environment. Binding of various factors serves as logic switches. TF's are transcription factors, E's are enzyme pools, and X's are metabolic pools. From [19] (open access)

18.4 The Physical Basis: Linking the Two

To give the discussion a specific context, I will now focus on the brain.

18.4.1 The Nervous System

Brains are based in the underlying physics through the operation of neurons linked by synapses and structured in neural networks, in particular forming layered columns in the neocortex [25].

Neurons receive spike trains via axons, which flow to the nucleus where a summation operation is performed, and resulting spike trains then flow down axons to synapses where a further summations process takes place; signals are passed to other neurons if the outcome is above a threshold [25]. The behaviour of currents in dendrites and axons is governed by the underlying physics (described by Eqs. (18.2–18.4)). Quantum mechanical interactions based on these forces underlies the existence of the structures of neurons and their component parts. The Hodgkin-Huxley equations [22] charactize ion and electron flows that underlie existence of action potential spike trains in neurons. These equations follow from the structure of axons and dendrites, and in particular from the the existence of ion channels [13] that allow ions to selectively flow in and out of the cell membranes. The constants in these equations are not universal constants, but are constants characterising the axon structure and environment. They cannot be deduced from the laws of physics alone [43].

18.4.2   The Link of Physics to Logic: The Molecular Basis

The logic of what happens is enabled by voltage gated ion channels in axon and dendrite membranes [13, 27]. They lead to controlled flow of sodium, potassium and chlorine ions into and out of the axons and dendrites. The ion channels implement logical operations with the following logic[4]:

$$\text{If voltage } V > V_0 \text{ then let ions flow, else not} \qquad (18.9)$$

which is of the form (18.5). It is the structural form of the ion channels—the detailed three-dimensional molecular configurations characterised as tertiary and quaternary structures—that enables this logic to emerge out of physics in the brain at the lower levels, and so underlie emerging logic at higher level [25]. The relevant structure is that of incredibly complex proteins imbedded in the cell wall.

Given the existence of the ion channels, Maxwell's equations (18.2, 18.3) together with the equations of motion for particles (18.4) underlie what happens. The flow of ions through ion channels is governed by these physical laws and so has the physics form (18.1).

The implication is that, at least in the brain,

Molecular basis of "IF ... THEN ... ELSE ...." logic in biology: Biomolecules perform logical operations because of their tertiary and quaternary structure.

This applies equally in many other biological contexts (see below), where they can be combined to give the basic operations AND, OR, NOT. These basic logical units are then combined via the specific neuronal connections and weights, in an almost incomprehensible way with with thousands of synaptic inputs on each neuron, leading to the coordinated actions involved in learning, memory, and higher level cognitive function. More complex logical operations occur, for example N-methyl-D-aspartate (NMDA) receptors are one class of receptors capable of responding to the high density of synaptic inputs on a single neuron. These receptors have a characteristic heterote-tramer between two NR1 and NR2 subunits that mediate numerous biological effects within neurons [29]. Most notable of these effects relate to learning and memory, particularly with respect to long-term potentiation (LTP) and synaptic plasticity in

---

[4]In practice, the response function is not discontinuous as in this representation, but is a smoother curve such as a logistic curve linking 'on' and 'off' states. The principle remains the same.

the hippocampus. Unlike typical voltage-gated ion channels previously mentioned, NMDA receptors respond to the coordinated input from many synapses, leading to global depolarization across the cell membrane before activating and allowing calcium entry into the cell [29]. Thus, NMDA receptors are linked to higher level functions in humans. The integrative nature of NMDA receptors illustrates the holistic nature of biological systems expressed by the late developmental biologist, Paul A. Weiss. In his book, *The Science of Life*, Weiss wrote [52],

> The complex [biological organism] is a system if the variance of the features of the whole collective is significantly less than the sum of variances of its constituents; or, written in a formula:
>
> $$V_s < (V_a + V_b + ... + V_n) \qquad (18.10)$$
>
> In short, the basic characteristic of a system is its essential invariance beyond the much more variant flux and fluctuations of its elements or constituents... This is exactly the opposite of a machine, in which the pattern of the product is simply the terminal end of a chain of rigorously predefined sequential operations of parts. In a system, the structure of the whole coordinates the play of the parts; in the machine, the operation of the parts determines the outcome. Of course, even the machine owes the coordinated functional arrangement of its parts, in last analysis, to a systems operation—that of the brain of its designer.

Similarly, there exists a dynamic interplay between single voltage-channel receptors and receptors responding to the global interplay of all synaptic inputs on postsynaptic neurons. Thus, the brain exists as a dynamic entity of intertwining top-down and bottom-up processes.

Energy is of course used in carrying out these logical processes, and local energy minimisation will occur as part of what is going on (e.g. in the protein folding that converts one-dimensional strings of amino acids to the very complex tertiary and quaternary structures of proteins [36]). But energy or entropy considerations will not by themselves produce the desired logical operations that enable the function and purpose characteristic of life [21] to emerge. Indeed the necessary energy usage for cellular function is controlled by complex metabolic regulatory networks [51] that determine what energy transactions will take place on the basis of logical operations; Fig. 18.2 gives a simple example, showing how they rely on basic logical operations (18.5) mediated by transcription factors that can be "on" or "of". Thermodynamics and statistical mechanics are not by themselves enough to capture what is going on.

In summary, *biomolecules such as ion channels* [13, 27] *can be used to make logic gates, and the higher level physiological systems in which they are imbedded* [12, 19, 25, 38] *enable emergence of complex life logic and processes.*

18.5   The Existence of the Ion Channels

Two issues arise here: the possibility of existence of the molecules that comprise ion channels, and how they come into being.

18.5.1   Their Possible Existence

The possible existence of biomolecules, and particularly the proteins that govern biological activity [36], results from quantum interactions mediated by the electromagnetic force [51]. Indeed both the existence of atomic nuclei and of molecular binding forces cannot be explained classically. But given the nature of physics as we know it, with the nature of everyday scale structures controlled by electromagnetism together with quantum physics, there is a resulting space of possible proteins of vast dimensions: a kind of Platonic space of possible structures [50]. Such a platonic space has been proposed to explain the common structural motifs found within all proteins in biological systems ([14]).

18.5.2   Their Coming into Being

Given this possibility space, how have the specific proteins that exist and control biological function come into being? This has developmental and evolutionary aspects.

Developmental aspects The relevant proteins exist because of the reading of the genetic information written into our DNA [2, 51] through developmental processes [16, 34, 53], resulting in amino acid chains that fold to form biologically active proteins. This reading of the genotype takes place in a contextual way [31] because of epigenetic processes [17].

Evolutionary aspects How did that genetic information came to exist? Equivalently, how did the specific proteins that actually exist [36] come be selected from all of those that might possibly have existed, as characterised by the vast space of protein possibilities [50]? These extraordinary complex molecules with specific biological functions (for example, hemoglobin exists in order to transport oxygen in our blood stream) cannot possibly have come into being simply through bottom-up self assembly. They have to have been selected for through the process of adaptive selection [12] acting on a pool of genes that have been varied due to mutations, random genetic drift, and horizontal gene transfer, with massive degeneracy in the genotype-phenotype map playing a crucial role in enabling new genotypes to come into being in the time available [49]. This selection process in essence results in new information being embodied in the sequences of base pairs in the DNA [51] that was not there before: in effect, this is writes information about the environment into genes [44]. Then reading out that information by cellular processes [51] creates the string of amino acids that forms proteins such as haemoglobin. It then has to fold to give its biologically active form. In principle that step is an energy minimisation operation; in practice it requires molecular chaperones to achieve the required folding. These are a further set of proteins have to be coded for in DNA and whose existence has to be explained on the basis of natural selection. Additional energy is also used for post-translational modification to produce the final protein structure.

18.5.3 The Basic Selection Process

The basic generic selection process is that a random input ensemble of entities is filtered to produce an ordered output ensemble, adapted to the environment via specific selection criteria. The logic of the process is a special case of that given in Eq. (18.5):

$$\Pi_S(X) : \{\text{IF } X \notin S(C, \mathcal{E}) \text{ THEN DELETE } X\} \quad (18.11)$$

where S is the subset of elements selected to survive if C is the selection criterion and $\mathcal{E}$ the environmental context. The effect on the ensemble $\{E(X)\}$ is a projection operation:

$$\Pi_S : \{E(X)\} \to \{\hat{E}(X) : X \in S(C, \mathcal{E})\}. \quad (18.12)$$

So now projecting again, Π_S leaves the new ensemble $\{\hat{E}(X)\}$ invariant:

$$\Pi_S : \{\hat{E}(X)\} \to \{\hat{E}(X)\}. \quad (18.13)$$

A simple example in the logical case is deleting emails or files on a computer; the basic physics case is Maxwell's Demon [48]. It takes place in biology through Darwinian selection [12], where the input ensemble at time t_2 is the output of the previous process at time t_1, randomised to some degree:

$$\{E(X)\}(t_2) = R\{\hat{E}(X)(t_1)\}. \quad (18.14)$$

where R is a randomisation operation based in mutations and genetic drift. It is the remorseless continual repetition of the process of variation (restricted by physiological, and development possibilities [30]) and subsequent selection that gives evolution its extraordinary creative power, underlying the emergence of complex life forms [12]. While much of this happens by alterations in DNA, additionally, growing evidence suggests that epigenetic mechanisms, such as DNA methylation, may also be involved in rapid adaptation to new environments [28, 37]. This is all constrained by physics: in essence, physical laws themselves provide an important contribution for direction and function of living systems. The overall process results in an increase of mutual information [44] between the system and the environment; equivalently, the organism has adapted so as to reduce surprisal [44] as it moves through its environment. This takes place either by the organism adapting its structure and behaviour to the environment, or by the organism altering that environment to suit its needs (niche construction). In the human case, that reshaping is achieved by technological means derived from the creative activity of the human mind [20].

18.5.4 A Multilevel Process

In biology, this process of adaptation takes place in a contextual way through evolutionary emergence of developmental systems at the lower levels [34] that require existence of specific proteins for their function. At the higher levels it leads to development of robust physiological systems [38] (protected by homeostasis) and a plastic brain that can adapt effectively to the physical, ecological, and social environment. An example is that vision gives great survival advantage to individuals. Development of vision is a multi-level process, with higher level needs driving lower level selection of structure and function:

- The top level need is for a visual system that will enhance survival;
- The next level need is for eyes, an optic tract, thalamus (a relay station for signals on the way to the cortex), and neocortex to analyse incoming data;
- The next level is a need for photo receptor cells within the retina, and neurons and synapses to constitute neural networks to analyse the data;
- One then needs specific kinds of proteins to make this all work [50], for example rhodopsins in Light Harvesting Complexes and voltage gated ion channel proteins in the neuronal axons;
- So one needs to select for developmental systems [34] to make this all happen, comprised of Gene Regulatory Networks, Signal Transductions networks, and Metabolic Regulatory Networks, together with the proteins needed to make them work [50];
- Thus one needs whatever genome will do the job of providing all the above [50].

This is all guided by high level needs, as made clear in this example. The environmental niches $\mathcal{E}$ might be that the animal lives on land, or in the air, or in shallow water, or in deep water. The selection criterion C might be a need to mazimise intensity sensitivity, or edge detection, or motion detection, or angular resolution, or colour sensitivity; which is most important will depend on the ecological environment. Thus natural selection is a top down process [15] adapting animals to their environment in suitable ways, thereby altering the details base-pair sequence in DNA. As stated by Stone ([44]:188):

> Evolution is essentially a process in which natural selection acts as a mechanism for transferring information from the environment to the collective genome of the species.

Actually it is doubly a top down process, through the environment creating niches $\mathcal{E}$ (opportunities for life) on the one hand, and through the selection criteria C on the other. Altering either of them alters the micro (genotype) and macro (phenotype) outcomes. The reliability of the resulting systems is because, as discussed above, biological systems consisting of hierarchically nested, complex networks that are extremely robust to extrinsic perturbations, which are the context within which these adaptive processes take place [47].

18.5.5   What is the Role of Chance?

There is a great deal of noise and randomness in biological processes, particularly at the lower levels where molecules live in what has been labelled a 'molecular storm' [23]. The occurrence of this noise does not mean the outcome is random: rather, it provides an ensemble of variants that is the basis for selection of outcomes according to higher level selection criteria, thus creating order out of disorder, as in (18.12). Indeed, microbiology thrives on randomness [23] as does brain function [18, 40]. Statistical randomness between levels provides the material on which selection processes can operate. Quantum uncertainty might also play a role (this is unclear until we understand the nature of the quantum measurement problem).

18.6   Conclusion

How do goal-oriented systems arise out of the goal-free underlying physics? The main conclusion reached here is that

> **Protein structure forms the link between physics and biology at the micro scale, as discussed above, underlying emergence of macro-scale purposive entities when incorporated in complex networks. The proteins and the network must both be selected for through processes of adaptive selection.**

We have used ion channels as our main example, because they underlie signal processing in the brain, but there are many other biomolecules that are used in interaction networks to carry out logical operations. In particular transcription factors binding to specific DNA sequences operating by the lock and key molecular recognition mechanism enable logical operations such as AND and NOT (Fig. 18.2), as do the operation of synaptic thresholds associated with excitatory or inhibitory receptors in neurons.

18.6.1   Adaptation and Plasticity

Life depends on adaptation to its environment. Adaptation takes place at all levels and timescales: evolutionary, developmental, and functional. Brain operations such as learning are context dependent and adaptive [5, 15]; so are operations of all physiological systems [31], and all molecular biology processes [17]. The brain plasticity at macrolevels that underlies our adaptive behaviour is enabled at micro levels by biomolecules acting as logical devices (18.5) choosing alternative outcomes on the basis of local and global variables. This is where the key difference from purely physically based interactions occurs. In biology, structure and function go hand in hand [12]; and this is in particular true in the relation between macro systems and their underlying molecular and cellular structure, which is a two-way process: influences are both bottom-up and top-down [15].

18.6.2   The Nature of Adaptation

The underlying physics enables adaptive selection processes to happen, but the physics does not by itself decide what will happen; this is driven by higher level needs (e.g. development of eyes, as discussed above). Adaptive selection is an emergent biological process. It acts at all levels: on groups, organisms, systems, cells, interaction networks, genes, and molecules [12]. It is based in physical processes, but is not itself a physical law: it is an essentially biological effect. It is not directly implied by or deducible from the equations of the standard model of particle physics. In summary,

Physics underlies adaptive selection in that it allows the relevant biological mechanisms to work; but adaptive selection is not a physical law. It is an emergent biological process.

Adaptive selection could not take place without physics, but unlike physics is a purposeful process in that it has the logic of increasing fitness. It is irreversible, because species die out in order that others succeed; logically this is because it follows from (18.12) that

$$\{\hat{E}(X)\} \, does \, not \, determine \, \{E(X)\}, \tag{18.15}$$

so information $\Delta E(X) := \{E(X)\} - \{\hat{E}(X)\}$ is lost in the selection process, unless $\{E(X)\} = \{\hat{E}(X)\}$. This process is influenced at higher levels of development by social and psychological influences that crucially shape outcomes [20], for example through the development of the language capacity that distinguishes us from the great apes: a symbolic capacity enhances survival because it enables the development of technology [11].

Adaptive selection is not the same as energy minimisation, although that will play an important part in determining what can happen, nor is it entropy maximisation. It cannot be deduced by statistical physics methods [35], nor by a consideration of force laws such as (18.2–18.4), or from the standard model of particle physics. It is not implied by physics, which has no concept of survival of a living being (or for that matter, of a living being), but is enabled by it.

The key thing that enabled this all to happen was the origin of life, when adaptive evolutionary processes came into being. We still do not know how that happened.

Appendix: Technical notes

1. This essay is a companion to a paper with Philippe Binder on the relation between computation and physical laws [7].
2. In parallel to the discussion here, digital computers also show a high degree of logical complexity based in the logical capacities built into low level devices (transistors) that are combined so as to create logic gates underlying a tower of virual machines (see [15]). It is these specific physical micro-structures, in this case products of logical analysis by the human mind, that enable the underlying physical laws to generate logical behaviour. Given this physical structure, computer programs specify the abstract logic that will be carried out in a particular application, e.g. word processing or image editing. The same physically based microstructure can carry out any logical operations specified [15].
3. The existence of bio-molecules is enabled by covalent bonds, hydrogen bonds, and van der Waals forces [51]. All are based in quantum physics and the electromagnetic interaction (18.2–18.4).
4. Multiple realisation of higher level functions and processes at lower levels is a key feature in the emergence of complexity such as life. The key analytic idea is that of identifying functional equivalence classes [2, 15]: each equivalence class is a set of lower level properties that all correspond to the same higher level structure or function. This degeneracy occurs in the way developmental systems are related to the genome: a vast number of different genomes (a *genotype network*) can create the same phenotype [50]; any one of them can be selected for and will do the job needed. This huge degeneracy solves the problem of how biologically effective alternatives can be explored in the time available since the start of life, as explained by Wagner [50] in his important book. These are what get selected for when adaptation takes place; and it is the huge size of these equivalence classes that enables adaptive selection to search out the needed biomolecules in the time available since the origin of life [50]. Whenever you can identify existence of such functional equivalence classes, that is an indication that top-down causation is taking place: see Auletta et al. [6].
5. In the case of a complex logical system, you do not get the higher level behaviour by coarse graining, as in the case of statistical physics [35]. You get it by black boxing and logical combination, involving information hiding and abstraction to characterize the exterior behaviour of a module, see Ross Ashby's book *An Introduction to Cybernetics* [5], and Giulio Tononi et al's work on Integrated Information Theory [33]. This is particularly clear in the case of digital computer systems, with their explicit apparatus of abstraction, information hiding, and carefully specified module interfaces, see Grady Booch's book *Object Oriented Analysis* [10].

References

1. Adler, J., Tso, W.-W.: 'Decision'-making in bacteria: chemotactic response of escherichia coli to conflicting stimuli. Science **184**, 1292–1294 (1974)
2. Alberts, B., Johnson, A., Lewis, J., Raff, M., Roberts, K., Walter, P.: Molecular Biology of the Cell. Garland Science (2007)
3. Arnol'd, V.I.: Mathematical Methods of Classical Mechanics. Springer (1989)
4. Ashby, W.R.: An Introduction to Cybernetics. Chapman and Hall, London (1956). http://pespmc1.vub.ac.be/ASHBBOOK.html
5. Ashby, W.R.: Design for a Brain: The origin of adaptive behaviour. Springer (1960) https://doi.org/10.1007/978-94-015-1320-3.epub
6. Auletta, G., Ellis, G., Jaeger, L.: Top-down causation: from a philosophical problem to a scientic research program. J R Soc. Interface B, 1159–1172(2008)
7. Binder, P.M., Ellis, G.F.R.: Nature, computation and complexity. Phys. Scr. **91**, 064004 (2016)
8. Binney, J., Tremaine, S.: Galactic Dynamics. Princeton (2008)
9. Blundell, S.J.: Katherine Blundell. Concepts in Thermal Physics. Oxford University Press, K.M. (2008)
10. Booch, G.: Object Oriented Analysis and Design with Applications. Addison Wesley, New York (1994)
11. Bronowski, J.: The Ascent of Man. BBC, London (1973)
12. Campbell, N.A., Reece, J.B.: Biology. Benjamin Cummings, San Francisco (2005)
13. Catterall, W.A.: From ionic currents to molecular mechanisms: the structure and function of voltage-gated sodium channels. Neuron **26**, 13–25 (2000)
14. Denton, M.J., Marshall, C.J., Legge, M.: The protein folds as platonic forms: new support for the pre-darwinian conception of evolution by natural law. J. theor. Biol. **219**, 325–342 (2002)
15. Ellis, G.F.R.: How Can Physics Underlie the Mind? Top-Down Causation in the Human Context. Springer (2016) web page
16. Gilbert, S.F.: Developmental Biology. Sinauer (2006)
17. Gilbert, S.F., Epel, D.: Ecological Developmental Biology. Sinauer (2009)
18. Glimcher, P.W.: Indeterminacy in brain and behaviour. Ann. Rev. Psychol. **56**, 25–56 (2005)
19. Goelzer, A. et al.: Reconstruction and analysis of the genetic and metabolic regulatory networks of the central metabolism of Bacillus subtilis. BMC Syst. Biol. **2**, 20 (Available here)
20. Harford, T.: Fift Things That Made the Modern Economy. Little Brown (2017)
21. Hartwell, L.H., Hopfield, J.J., Leibler, S., Murray, A.W.: From molecular to modular cell biology. Nature **402**(Supplement): C47–C52, pdf here
22. Hodgkin, A.L., Huxley, A.F.: A quantitative description of membrane current and its application to conduction and excitation in nerve. J. Physiol. **117**, 500–544 (1952)
23. Hoffmann, P.M.: Life's Ratchets: How Molecular Machines Extract Order from Chaos. Basic Books, New York (2012)
24. Janes, K.A., Yaffe, M.B.: Data-driven modelling of signal-transduction networks. Nat. Rev. Mol. Cell Biol. **7**, 820–828 (2006)
25. Kandel, E., Schwartz, J.H., Jessell, T.M., Siegelbaum, S.A., Hudspeth, A.J.: Principles of Neural Science. McGraw Hill Professional (2013)
26. Krebs, H.: The citric acid cycle. Nobel Lecture, 11 Dec 1953
27. Magleby, K.L.: Structural biology: Ion-channel mechanisms revealed. Nature **541**, 33–34 (2017)
28. McNew, S.M., Beck, D., Sadler-Riggleman, I., Knutie, S.A., Koop, J.A.H., Clayton, D.H., Skinner, M.K.: Epigenetic variation between urban and rural populations of Darwin's finches. BMC Evol. Biol. **17**, 183 (2017)
29. Nestler, E.J., Hyman, S.E., Holtzman, D.M.: Molecular Neuropharmacology: A Foundation for Clinical Neuroscience, 3rd edn. McGraw-Hill, New York (2015)
30. Noble, D.: Neo-darwinism, the modern synthesis and selfish genes: are they of use in physiology? J. Physiol. **589**, 1007–015 (2011)

31. Noble, D.: A theory of biological relativity: no privileged level of causation. Interface Focus **2**, 55–64 (2012)
32. Noble, D.: Dance to the Tune of Life: Biological Relativity. Cambridge University Press (2016)
33. Oizumi, M., Albantakis, L., Tononi, G.: From the phenomenology to the mechanisms of consciousness: integrated information theory 3.0. PLoS Comput. Biol. **10**, e1003588 (2014)
34. Oyama, S., Griffiths, P.E., Gray, R.D.: Cycles of Contingency: Developmental Systems and Evolution. MIT Press, Cambridge Mass (2001)
35. Penrose, O.: Foundations of statistical mechanics. Rep. Prog. Phys. **42**, 1937–2006 (1979)
36. Petsko, G.A., Ringe, D.: Protein Structure and Function. Oxford University Press, Oxford (2009)
37. Pigliucci, M., Müller, G.B.: Evolution–the Extended Synthesis. MIT Press, Cambridge Mass (2000)
38. Rhoades, R., Pflanzer, R.: Human Physiology. Saunders College Publishing, Fort Worth (1989)
39. Riley, J., Greggers, U., Smith, A., Reynolds, D., Menzel, R.: The flight paths of honeybees recruited by the waggle dance. Nature **435**, 205–207 (2005)
40. Rolls, E.T., Deco, G.: The Noisy Brain: Stochastic Dynamics as a Principle of Brain Function. Oxford University Press, Oxford (2010)
41. Sakai, T., Haga, K.: Molecular genetic analysis of phototropism in arabidopsis. Plant Cell Physiol. **53**, 1517–1534 (2012)
42. Shockley, W.: The theory of p-n junctions in semiconductors and p-n junction transistors. Bell Labs Tech. J. **28**, 435–489 (1949)
43. Scott, A.: Stairway to the Mind The Controversial New Science of Consciousness. Springer (1995)
44. Stone, J.V.: Information Theory: A Tutorial Introduction. Sebtel Press (2015)
45. Taff, L.G.: Celestial mechanics: A computational guide for the practitioner. Wiley-Interscience, New York (1985)
46. Tanenbaum, A.S.: Structured Computer Organisation. Prentice Hall, Englewood Cliffs (2006)
47. Tëmkin, I., Eldredge, N.: Networks and hierarchies: approaching complexity in evolutionary theory. In: Serrelli, E., Gontier, N. (eds.) Macroevolution. Springer (2015)
48. Von Baeyer, H.C.: Maxwell's Demon: Why Warmth Disperses and Time Passes. Random House (1998)
49. Wagner, A.: The Origins of Evolutionary Innovations. Oxford University Press, Oxford (2011)
50. Wagner, A.: Arrival of the Fittest. Penguin Random House (2017)
51. Watson, J.D., Bell, S.P., Gann, A., Levine, M., Losick, R., Baker, T.A.: Molecular Biology of the Gene. Pearson (2013)
52. Weiss, P.A.: The Science of Life: The Living System–A System for Living. Futura Pub. Co., Mount Kisco, NY (1978)
53. Wolpert, L.: Principles of Development. Oxford University Press (2002)

Appendix
List of Winners

First Prizes[1]

Larissa Albantakis: A Tale of Two Animats: What does it take to have goals?
Carlo Rovelli: Meaning and Intentionality = Information + Evolution
Jochen Szangolies: Von Neumann Minds: A Toy Model of Meaning in a Natural World

Second Prizes

No second prizes were awarded

Third Prizes

Simon DeDeo: Origin Gaps and the Eternal Sunshine of the Second-Order Pendulum
Erik P Hoel: Agent Above, Atom Below: How agents causally emerge from their underlying microphysics
Dean Rickles: World without World: Observer-Dependent Physics
Ines Samengo: The role of the observer in goal-directed behavior
Rick Searle: From Athena to AI: the past and future of intention in nature
Marc Séguin: Wandering Towards Physics: Participatory Realism and the Co-Emergence of Lawfulness

Fourth Prizes

Tommaso Bolognesi: The Man in a Tailcoat
Ian Durham: God's Dice and Einstein's Solids
Alan M. Kadin: No Ghost in the Machine
Sophia Magnusdottir: I think, therefore I think you think I am

---

[1]From the Foundational Questions Institute website: http://fqxi.org/community/essay/winners/2016.2.

A. Aguirre et al. (eds.), *Wandering Towards a Goal*, The Frontiers Collection,
https://doi.org/10.1007/978-3-319-75726-1

Cristinel Stoica: The Tablet of the Metalaw
Sara Imari Walker: Bio from Bit
Noson S. Yanofsky: Finding Structure in Science and Mathematics

Special Community Prize

George F. R. Ellis: Wandering Towards a Goal: The Key Role of Biomolecules

Titles in This Series

Quantum Mechanics and Gravity
By Mendel Sachs

Quantum-Classical Correspondence
Dynamical Quantization and the Classical Limit
By A.O. Bolivar

Knowledge and the World: Challenges Beyond the Science Wars
Ed. by M. Carrier, J. Roggenhofer, G. Kuppers and P. Blanchard

Quantum-Classical Analogies
By Daniela Dragoman and Mircea Dragoman

Quo Vadis Quantum Mechanics?
Ed. by Avshalom C. Elitzur, Shahar Dolev and Nancy Kolenda

Information and Its Role in Nature
By Juan G. Roederer

Extreme Events in Nature and Society
Ed. by Sergio Albeverio, Volker Jentsch and Holger Kantz

The Thermodynamic Machinery of Life
By Michal Kurzynski

© Springer International Publishing AG, part of Springer Nature 2018
A. Aguirre et al. (eds.), *Wandering Towards a Goal*, The Frontiers Collection,
https://doi.org/10.1007/978-3-319-75726-1

Mindful Universe
Quantum Mechanics and the Participating Observer
By Henry P. Stapp

Principles of Evolution
From the Planck Epoch to Complex Multicellular Life
Ed. by Hildegard Meyer-Ortmanns and Stefan Thurner

The Second Law of Economics
Energy, Entropy, and the Origins of Wealth
By Reiner Kummel

States of Consciousness
Experimental Insights into Meditation, Waking, Sleep and Dreams
Ed. by Dean Cvetkovic and Irena Cosic

Elegance and Enigma
The Quantum Interviews
Ed. by Maximilian Schlosshauer

Humans on Earth
From Origins to Possible Futures
By Filipe Duarte Santos

Evolution 2.0
Implications of Darwinism in Philosophy and the Social and Natural Sciences
Ed. by Martin Brinkworth and Friedel Weinert

Probability in Physics
Ed. by Yemima Ben-Menahem and Meir Hemmo

Chips 2020
A Guide to the Future of Nanoelectronics
Ed. by Bernd Hoefflinger

From the Web to the Grid and Beyond
Computing Paradigms Driven by High-Energy Physics
Ed. by Rene Brun, Frederico Carminati and Giuliana Galli-Carminati

The Language Phenomenon
Human Communication from Milliseconds to Millennia
Ed. by P.-M. Binder and K. Smith

The Dual Nature of Life
Interplay of the Individual and the Genome
By Gennadiy Zhegunov

Natural Fabrications
Science, Emergence and Consciousness
By William Seager

Ultimate Horizons
Probing the Limits of the Universe
By Helmut Satz

Physics, Nature and Society
A Guide to Order and Complexity in Our World
By Joaquin Marro

Extraterrestrial Altruism
Evolution and Ethics in the Cosmos
Ed. by Douglas A. Vakoch

The Beginning and the End
The Meaning of Life in a Cosmological Perspective
By Clement Vidal

A Brief History of String Theory
From Dual Models to M-Theory
By Dean Rickles

Singularity Hypotheses
A Scientific and Philosophical Assessment
Ed. by Amnon H. Eden, James H. Moor, Johnny H. Søraker and Eric Steinhart

Why More Is Different
Philosophical Issues in Condensed Matter Physics and Complex Systems
Ed. by Brigitte Falkenburg and Margaret Morrison

Questioning the Foundations of Physics
Which of Our Fundamental Assumptions Are Wrong?
Ed. by Anthony Aguirre, Brendan Foster and Zeeya Merali

It From Bit or Bit From It?
On Physics and Information
Ed. by Anthony Aguirre, Brendan Foster and Zeeya Merali

How Should Humanity Steer the Future?
Ed. by Anthony Aguirre, Brendan Foster and Zeeya Merali

Trick or Truth?
The Mysterious Connection Between Physics and Mathematics
Ed. by Anthony Aguirre, Brendan Foster and Zeeya Merali

The Challenge of Chance
A Multidisciplinary Approach from Science and the Humanities
Ed. by Klaas Landsman, Ellen van Wolde

Quantum [Un]Speakables II
Half a Century of Bell's Theorem
Ed. by Reinhold Bertlmann, Anton Zeilinger

Energy, Complexity and Wealth Maximization
Ed. by Robert Ayres

The Seneca Effect
Why Growth is Slow but Collapse is Rapid
By Ugo Bardi

Chemical Complexity
Self-Organization Processes in Molecular Systems
By Alexander S. Mikhailov, Gerhard Ertl

The Essential Tension
Competition, Cooperation and Multilevel Selection in Evolution
By Sonya Bahar

The Computability of the World
How Far Can Science Take Us?
By Bernd-Olaf Kuppers

The Map and the Territory
Exploring the Foundations of Science, Thought and Reality
By Shyam Wuppuluri, Francisco A. Doria

Wandering Towards a Goal
How Can Mindless Mathematical Laws Give Rise to Aims and Intention?
Ed. by Anthony Aguirre, Brendan Foster and Zeeya Merali

Printed in the United States
By Bookmasters